MOTHER MEDIA

MOTHER MEDIA

Hot and Cool Parenting in the Twentieth Century

HANNAH ZEAVIN

The MIT Press
Cambridge, Massachusetts
London, England

The MIT Press
Massachusetts Institute of Technology
77 Massachusetts Avenue, Cambridge, MA 02139
mitpress.mit.edu

© 2025 Hannah Zeavin

All rights reserved. No part of this book may be used to train artificial intelligence systems or reproduced in any form by any electronic or mechanical means (including photocopying, recording, or information storage and retrieval) without permission in writing from the publisher.

The MIT Press would like to thank the anonymous peer reviewers who provided comments on drafts of this book. The generous work of academic experts is essential for establishing the authority and quality of our publications. We acknowledge with gratitude the contributions of these otherwise uncredited readers.

This book was set in Adobe Garamond Pro by New Best-set Typesetters Ltd. Printed and bound in the United States of America.

Library of Congress Cataloging-in-Publication Data is available.

ISBN: 978-0-262-04955-9

10 9 8 7 6 5 4 3 2 1

EU product safety and compliance information contact is: mitp-eu-gpsr@mit.edu

For my mother, Lynne Zeavin
For my son, Malachai Zeavin O'Brien

Like all forms of mediation between the biological individual, the atomic individual, and the integral society, the family is also deprived of its substance by the latter, similar to the economic sphere of circulation, or the category of education, which is deeply connected to the family. As a category of mediation, which in truth, even if without being aware of it, often only brought about the business of the entire totality, the family, apart from its eminent function, always had something illusory about it.
—Theodor Adorno, "On the Problem of the Family," 1955

Mothers of America
    let your kids go to the movies!
—Frank O'Hara, "Ave Maria," 1964

My boy and I rely on media to keep me strong enough to love.
—Simone White, "Or, On Being the Other Woman," 2022

Contents

INTRODUCTION: HOW MOTHER BECAME A MEDIUM *1*

PART I: THE MEDIA OF MOTHERING

1   **THE FIRST TECHNOLOGY IS HUMAN** *35*

2   **OUT OF THE CRADLE** *77*

PART II: PREDICTIVE MOTHERING

3   **HOT AND COOL MOTHERS** *113*

4   **SCREENING MOTHER, CODING BABY** *145*

PART III: THE MOTHERING OF MEDIA

5   **A HISTORY OF SCREEN PARKING** *177*

6   **FUTURE, TOMORROW, DREAM, SMART** *215*

    CODA: ANALOG KIDS *237*

Acknowledgments *245*
Notes *249*
Index *293*

Figure 0.1
Al Parker, illustration for Ray Bradbury's "The World the Children Made," *Saturday Evening Post*, September 23, 1950.

INTRODUCTION: HOW MOTHER BECAME A MEDIUM

". . . I wish you'd look at the nursery."[1]

So begins "The World the Children Made," Ray Bradbury's 1950 short work of speculative fiction. The story, which would be canonized under the title "The Veldt," imagined a future not unlike our present, in which homes run themselves and their inhabitants and the basic tasks of keeping a family alive are ceded to a complex series of automated forces beyond its control. Bradbury tells the tale of a particular mother and father, Lydia and George Hadley, who have grown ambivalent about their purchase of this highly technologized house—a brand-new "Happylife Home." On one occasion, Lydia remarks to George, "That's just it. I feel like I don't belong here. The house is *wife and mother now, and nursemaid*. . . . Can I give a bath and scrub the children as efficiently or quickly as the automatic scrub bath can? I cannot" (my emphasis).[2] While she previously had the assistance of a nursemaid, the house has taken over that position, and hers too. No longer an employer, no longer taken up with the toil of her tasks, Lydia, it seems, is now at loose ends—even if she feels she should enjoy getting free of what tied her down. Instead, the very thing supposed to liberate her—from the work of management and the work of care—renders her guilty, lesser, and obsolete. Everything had changed psychologically, socially, and politically—because something else runs the laundry, tends the children. Yet she cannot enjoy this freedom. She no longer has any function because previously her life's work was just that: a series of feminized functions we call care. Instead of a mother and nurse working together, there is now a single, unified maternal environment, what we would now call a smart home.

If the mother is in mourning about no longer being one, the twins, Wendy and Peter—recalling the names of other fictional children who once escaped a decidedly dull nursery and *their* nonhuman, canine caregiver for Neverland—are delighted by the endless entertainment the new home provides. The Happylife Home contains an augmented reality playroom, which serves as nanny, mother, psychotherapist, and television.[3] The room is comprised of three screen walls, and it shifts its form in keeping with whatever the children imagine. The children then interact with their fantasies in the absence of mediating parental supervision, as if an iPad without parental controls had become a room of the home, with a key difference: the children project their unconscious wishes onto it.[4]

This form of motherless engagement begets, on Bradbury's telling, a new illness: technological addiction, and its concomitant, alienation. Rather than curing them of their neuroses, the children are figured as downright antisocial. They are transfixed. Nothing comes between them and the playroom. The parents call in a Freudian psychologist (1950 was the heyday of Freud mania in the United States) who sees the use of augmented reality as a lapse in parental judgment, to say the least. The shrink is against the whole setup, perhaps because he feels himself becoming obsolete just as Lydia has: he should be reparenting the children, not a *house*. He claims there is only one way forward for the family: disconnect the house, reconnect the family. The family would be restored to its natural, unmediated order. Before the parents can flip the kill switch, the children trap their parents in the playroom: the virtual space of care becomes a real threat to family members and the family structure itself. The story concludes in a consequential crossing of the virtual and actual that is the opposite of a "happy life": the children are described as eating lunch while simultaneously two lions from the digital plains are enjoying their own meal—Lydia and George. The fantasy is realized, and not just on screen. The human mother has been eaten by its automated replacement.

Parents get eaten and their children go wrong when they dare mediate and remediate themselves. That is the concrete lesson of this short tale: do not be tempted by devicing, it will do you in and rear, in your stead, a psychopathic generation. If Bradbury's 1950s readers took this death by glitchy digital lion as metaphor, they might have read the story as cautioning that no form of entertainment, no form of automation, was worth its risks. Suddenly, in 1950, this was evermore an option, purchasable by the middle class, if only in the form of a refrigerator and television, rather than an automated playroom. Bradbury offered a techno-pessimist story from the moment in which many middle-class homes were being remade, as were the families who lived in them. Morally and socially, Bradbury affirmed the nuclear family as the antidote to over-mediation even as he described its susceptibility to it. He offered a fantasy that—*pace* the psychoanalyst—mother and father could and should be a world unto themselves. Or else.

The family given to us by Bradbury is a fiction, but so is the nostalgic image of the family that circulates today. Historians of the US family have long worked against the grain of its popular imaginings, its mythologies, its outsized attention to the nuclear family at mid-century, which do the work of standing in for *all* families, and at all times. The idea of mother as a medium of care is rooted in the idealized Victorian bourgeois family and relates to a longer history of the detached home and, with it, what Raymond Williams calls "mobile privatization."[5] Yet, as readers bemoaned the fate of Lydia and George, family life was far from immutable: it was changing rapidly, and it would change again. The image of the family at mid-century remains, however, *culturally* arrested, fixed like the photographs and advertisements from that moment that remain in circulation or have been updated only to rearrest us now. The family that this image calls to mind was, in the words of the historian of the US family Stephanie Coontz, "atypical," and American mothering in this period was, in no uncertain terms, exceptional.[6] She writes, "For the first time in 80 years, the age of marriage fell sharply, fertility rates increased, and the proportion of never-married individuals plummeted. The values attached to nuclear-family living, including the rejection of 'interference' by extended kin and the expectation that family life should be people's main source of personal gratification, were also new—and their hegemony even at the time should not be exaggerated."[7]

We will return to the question of how a vision of the family not at all representative in its own time—let alone ours—became representative of a nation. But first, Coontz' allusion to "interference" points to a new conception of family self-determination and self-securing. Coontz argued that the older forms of multigenerational family life were newly understood as impinging upon young families—and that these families should be self-sufficient both in terms of emotional fulfillment and in terms of labor. The multigenerational household was under suspicion. After all, part of the selling point of the suburban home was precisely that young families could "get away" from their extended kinship networks to make family alone. The waged care that often supported middle- and upper-class family life was also newly suspicious. This auxiliary or dispersed care was in decline as workers sought new forms of waged opportunity. These older forms of care, we might say, were understood *as* interference in the family in women's magazines, in parenting literature, and even in domestic policy. How members of the new nuclear family should communicate with one another, and do so directly, without mediation, especially in maternal-child relations, was the site of extensive research across an astonishing array of social sciences: anthropology, sociology, psychology, psychoanalysis, and psychiatry, as well as education and medicine.

A new consternation emerged about domestic technologies, from household appliances to media technologies—and the media they conveyed—and how they affected the American family. Not only were the intimate relations between mother and child a site of concern, but the nuclear family was also under scrutiny for how it managed (and mismanaged) communications and transmissions coming from beyond its walls. Just as media forms were understood as mother proxies (the kernel of social realism in the science fiction of "The Veldt"), metaphors for media were newly employed to describe, classify, and often pathologize the mother. While Bradbury portrayed a white middle-class family, psychologists scrutinized mothers as a category, wholesale, articulating new theories about communication, bonding, and the import of maternal labor. Mothering was dissected and redescribed via class, race, gender, and sexuality, generating a new idiom for the criminalizing and pathologizing of mothers along these same axes. In this same moment—the middle of the twentieth century—new media forms such as television (and novels, comic books, and cinema) were understood as occupying children's time in lieu of adequate human care. How children responded both to mother and to media was newly top of mind for a wide class of social science researchers.

As is apparent in "The Veldt," psychological and technical discourses were seemingly newly entangled in older forms of care and social reproduction—often in opposition to one another. On the one hand, mothers were negotiating attempts by architects and industrial designers to automate their labor, and on the other hand, social scientists and medical professionals told them they needed to be exquisitely attentive and "always on" in the home—they were told to be as available, as infallible as the media they were supposed to avoid—to be mothering machines.

Family mediation—how the family negotiates society, including new media—became the definitional crisis of the family in the twentieth century. At the center of it was mother. This crisis might feel familiar to contemporary readers. For the idealization of good mothers and the pathologization of bad mothers (and parents) continues to circulate, among other panics, as one of maternal mediation and of media. This book, therefore, tells the story of how our contemporary understanding of a mother came to be—a figure encumbered by sociopolitical deputization and assistive devices. This profound shift has largely escaped notice, for it has been naturalized. In every concerned glance, chiding "tsk," or sigh directed at a mother shushing her child while handing over an iPad, the quiet development of supererogatory judgment of mothers has been hidden from cultural and historical view.

Mother Media tells the story of this reconfiguration of the mother: how, in the United States, a set of experts laid the ground for conflating mothers with media, passing the pathologies each supposedly wrought across categories. Starting in the 1930s, there was new and exquisite attention to "nuclear family living" in the human sciences, which studied and diagnosed this image of the family and offered new and long-standing definitions and understandings of what a mother is, what a father is, and what a child is. These definitions have a long tail and haunt us to this day.

We know that at mid-century there were intense political, social, moral, and economic pressures on mothering. Similarly, we know that there were intense pressures on media—in their standardization, in their domestication, in their promoted uses, and in their curbed misuses. For much of the twentieth century, two bodies of research have focused on the mother and media, together and apart, as the site of familial illness: the psy-ences (particularly psychiatry and psychoanalysis) and, perhaps more surprisingly, the nascent category of media researcher. *Mother Media* shows how these social panics were so linked as to be two sides of the same coin, and it offers us a new account of the influence of psychological theories on both mothering and media—how they meet one another in the living room—and on family life in the twentieth-century United States. I argue that pathologies of mothering became attached to pathologies of media—intensifying from 1940 to 1980 in particular—and vice versa. The book, therefore, offers a double genealogy, one tracing the emergence of what I call a "parenting theory of media" put forward by those working on media and a "media theory of mothering" put forward by psychologists working on mothering. Together, they became a definitional heuristic by which mothers were understood psychosocially in the mid-century and to our present day. The two theories united mothering and media to form a strange and strained slippage across media (TV and comic books, but also baby monitors and technologized cribs), mediation (a process by which two discrete terms are brought into relation; for media theorists, often by the medium of communication), and medium (as the *Oxford English Dictionary* has it, "the intervening substance through which impressions are conveyed"). Looking at the whole, this was a complex and dynamic mediation: media, the mother of all mediums; mother, the medium of all media.

*

Bradbury's story contains and expresses this interesting slip. He portrays the American mother as preferable to her wholesale remediation by automated technological means (in reality, this was much more partial, an extension, a supplementation). In Bradbury's moment, there was an efflorescence of psychological theories that thought of the mother as causing mass social degradation—as "evidenced" by a whole host of psychological studies conducted at the outset of World War II. American men were, therein, shown to be weak and psychoneurotic, and American mothers, psychologists argued, must have been to blame. But their automated replacements were no better—in fact, they too yielded, on Bradbury's telling, the destruction of the nuclear family and, with it, society. As I will argue, for psychologists, mothers posed a problem because they *mediated* the child, and on that understanding, mothers did so to disastrous result. The remediation of mothers was a problem because, in turn, her technological counterparts mediated children improperly—influencing them and supplanting the mother. Whereas mothers were considered the better alternative to those standing in for them, the mother was a fallible medium of care herself. Psychologists thus had to negotiate this cognitive dissonance of categorizing the mother as pure, on the one hand, and as a corrupting influence, on the other. They did so by making a whole new set of diagnoses of both mother and child that recorded the distance between an ideal of American mothering and its actualities.

By mid-century, we can understand the crisis of the family as one of mediation. The left and the right, as we will see, often took up the same work produced by social scientists about the home to make claims that rhymed with the pervasive attitudes of familial purity (what I will term *anti-mediation*) implied by the very objects they critiqued. As historian John Demos puts it, "broadly speaking, the history of the family has been a history of contraction and withdrawal; its central theme is the gradual surrender to other institutions of functions that once lay very much within the realm of family responsibility."[8] This book will look at the family's surrender—but also its resistance—to pathologization.

As we shall see, there is a longer story to tell about the conflation of mothering and the mediation of the child. But the interwar period and World War II set into motion the conditions for temporary psychological laboratories (prisons, orphanages, hospitals) that were instructive in the study of child-mother relations, especially via their separation. The dislocation of mother and child allowed researchers to move evermore from theoretical children to real children, both in the usual consulting room and nursery and beyond. Research attached to new family-in-crisis formations: for instance, absent fathers at the front in the war, children undergoing displacement, an increase in orphans. Taken together, there was a veritable eruption in psychological research that empirically documented, theorized, and studied the impacts of mother: what kinds of care she offered, its ratios of presence and absence, the quantities of attention, and the qualities of her connection.

These clinical findings in turn produced new dicta about mothering, thinking of her love in terms of new developments in technology and media.

She was metaphorized via temperature (hot, cold) or as an atmosphere (an environment, a climate) that she regulated, conditioned. She became a space of enclosure: "enveloping," "holding," "securing," "containing." The massive attribution of importance to mother was inseparable from an accusation of her fallibility and its consequences. She became a matrix, a medium: the substance through which children acquired their health, morals, affects, ideologies, and psychological well-being.

To tell this story requires us to turn to the two groups of thinkers most absorbed in the problem and most public in their commentary—a wide set of child researchers and a wide set of media researchers. In their experiments, archives, letters, and public conclusions, an overwhelming number of psychologists, psychoanalysts, and pediatricians determined nothing less than a media theory of mothering. This is not to say that they conceived of themselves as evincing this theory, but that we can see time and again that the mother is configured as a medium of care, of communication, and that pathological outcomes in children were tied to pathological mother-mediums.[9] In parallel, media gurus, theorists, critics, and evangelists began to understand media as having the capacity—for good or for ill—to do the work of standing in for mother. Following thinkers who have offered us deft histories of media research, such as Anna Schectman, John Durham Peters, Fred Turner, Lynn Spigel, and Bernard Geoghegan, we know that what we now might call "media theory" intersected with the remaking of the human and social sciences. We know from them that starting with the postwar era, and cresting through the late 1960s, theories of communication—including familial communication—were effervescent, and new departments and schools of communication were founded. Subsequently, as we will see, media studies was consecrated as its own field, particularly around the advent of media ecology (see more below). To these accounts of these fields, I add here that particular attention to media (both its medium and its content) converged with new ways of understanding psychological etiology in two key cultural and social assertions: that the success of child-rearing redounded to mother, the all-powerful mediator of the family, and that mother, at the peril of her family and society, replaced herself with media.

Mother Media focuses on how those working in child development (both in pediatrics and in the psychological sciences) and technology both posed a limited definition of family norms by reconfiguring our understandings of domestic and mothering infrastructures and media relations simultaneously. This dual reconception of mothering and media can be traced all along the twentieth century: starting with the founding of the Children's Welfare Bureau in 1912 and the first smart home in 1915, when the drop in family size was already underway and consumerism was on the rise to our very present. Although chapters glance backward to the late nineteenth and early twentieth century, when meanings around childhood and family life begin to stabilize in the United States, the book focuses on its years of mother-as-media's most intense emergence and elaboration, roughly 1939–1980. Across the 1940s and 1950s, the human sciences—with a new focus on mother-infant interaction—figured the mother as central to upbringing, and they turned their attention from the pathological child to the normal one, then back to the pathological child again. In this mid-century moment, family ideals became ever more linked to the nuclear form (the term itself was coined in the 1920s), and the respectability of the working-class family became linked to the fantasy of the housewife. This is the stage upon which the increasingly powerful and professionalized psy-fields contributed new ways of diagnosing the failures and proprieties of familial management, sometimes appearing race-neutral (as in the diagnosis of white familial dysfunction) and otherwise overtly reifying racialized understandings of "proper" family and "fit" motherhood. The attention to mother as site of contamination of the next generation—as a medium—was tangled with the very devices in her home upon which her mothering might depend. Yet, before media could mother—and thus replace mother—mother had to be thought of as something of a medium herself.

FROM MATERNAL IMPRESSION TO MATERNAL MEDIUM

The word "mother," coming from *mater*, has always been material, and the mother-as-media paradigm did have precedents long before the twentieth century. Theories of maternal contagion have gone by many names—maternal imagination, mother's marks, maternal imprinting. In the early modern British context, historian Mary Fissell shows that there was thought to be not just a physical connection between the mother's and fetus's material bodies—an early "nature" argument—by which she might pass characteristics and ailments to her offspring, but a telepathic link. Early modern medical manuals understood the mother basically as a psychic inscription machine.[10] Her transmissibility was understood to be total, and in this sense, mothers have always been thought of as media—the literal medium through whom the "message" of the child had to pass into life. If she ate, thought, or did the wrong thing, it would be recorded in and on her developing child.

Forms of this belief survived the early modern period; for the next 400 years, a mother's thoughts; her milk and the milk of others, especially enslaved wet nurses; and her activities were all thought of as adversely impacting—mediating, shaping—but also nurturing her child.[11] These worries were not just in the white US imagination; they circulated in Indigenous communities—pre- and post-colonization—in what is now the United States, as well as in enslaved communities. Crucially, "maternal impression" was also a name given to panics about interacting across races (centrally, again, in wet-nursing).[12] This formed a quotidian conception of mothering as well as the backbone of the science of mothering, incorporated in daily life decisions around feeding and childminding. Maternal impression, too, might appear in spectacular cases, held out as a warning for all mothers—such as that of the Elephant Man, whose face was understood to be evidence of maternal impression of an elephant during pregnancy.[13]

At the end of the nineteenth century, as scholars like Viviana Zelizer and Paula Fass show, the United States and its families were transformed by new, strong, sentimental attachment to and, crucially, investment in children. In this period, nascent forms of child psychology were consecrated in the United States by John Dewey and James Mark Baldwin. Their early, and opposing, theories of child development were visions of the child as naturally progressive—irrespective of inputs or a nurture-forward environmental theory—in which the child was highly malleable or plastic (sometimes under a neo-Lamarckian conception). Some theories ventured an awkward admixture of the two antonyms.[14] If in the former conception, mother was medium because she carried and delivered the baby, in the latter, she both regulated the baby, mediating it, and formed it via impression. In nurture theories, mother (and school) was central to what became of child; under nature theories she might have a rather diminished status (which some women celebrated; see chapter 2).

By the first decades of the twentieth century until the late 1930s, pediatrics transformed the folklore of mother's marks into science by relying on the same cause and effect patterns articulated in the idiom of scientific mothering. On this model, mothers must *learn* the science of mothering and take full responsibility for the health of their child by using disseminated, expert knowledge. Scientific mothering, as we will see, laid the groundwork for a theory of the family that redounded to mother. As historian Rima D. Apple has it, under protocols of scientific mothering, "Women were both responsible for their families and incapable of that responsibility."[15] Scientific mothering carried with it a new surveillance of babies and mothers by doctors in the clinic, as well as rigid approaches to the tempo of mothering and infant life in the home. In the interwar period, these new sciences of child-rearing began to conceptualize threats to children. From emasculation to delinquency, all supposed characterological defects might come from something like maternal impression. More and more, these theories were narrow, and understood maternal impact as having a limited range of effects, which could be put bluntly: bad mothers produced bad children.[16] Behaviorists, most signally John Watson, thought of mothers as merely one input into children, who were outputs. If you input kissing to a child, say, you make that child, over time, "an invalid."

Between Watson's paradigm-making bible of psychological mothering, *Psychological Care of Infant and Child*, in 1928, and Dr. Benjamin Spock's monument to libidinal mothering, *The Common Sense Book of Baby and Child Care*, in 1946, the fields of child development, child psychiatry, and child psychoanalysis (often overlapping, and yet distinct) all would emerge. The various roots and trajectories may at first seem different, but many of the actors passed through one another's institutions and, in their cohorts, were largely responding to the same social forces and cultural conceptions that made diagnosing mothering central to pediatric etiology.

Important sites for the study of children and their mothers that populate this book include Yale University's Clinic of Child Development, Johns Hopkins's Phipps Clinic and Harriet Lane House at the children's psychiatry service, a constellation of psychoanalytically oriented clinics and medical centers in New York City (at Mount Sinai Hospital, Bellevue Psychiatric, and in Harlem at the Lafargue Clinic), and a New York prison where psychoanalysts studied mother-infant interaction. While some of these institutions or research units were helmed by those born in the United States, the impact of psychoanalytic training cannot be understated, which occurred in three ways: first in early US Freudian circles, especially after Freud's 1909 visit to the United States (he hated the country); second, with many US-born psychiatrists and lay would-be analysts studying in psychoanalytic centers in Europe, particularly Vienna and later Berlin; and finally, with emigration from Central and Eastern Europe to the United States before, during, and after World War II.

The field of child development became increasingly professionalized in this period, most notably under Arnold Gesell, who established the Clinic of Child Development at Yale University (without yet having earned his MD, which he did at his own institution). After earning his doctorate in psychology at Clark University just a few years before Sigmund Freud lectured there on his one American tour, Gesell founded the Clinic of Child Development in 1911. There, he would go on to incorporate one-way mirrors for observations, photographs, and films to document developmental reaction to stimuli, which became the basis for growth tables and intelligence tests.[17] He was also the author of several central works, including *An Atlas of Infant Behavior* (1934) and two well-selling pediatric child-rearing books, *Infant and Child in the Culture of Today* (1943) and *The Child from Five to Ten* (1946), the latter of which was outshone by the arrival of a work by Gesell's own student, Dr. Benjamin Spock, and his best-selling *The Common Sense Book of Baby and Child Care*, published in the same year.

Johns Hopkins, too, began its programs in child psychiatry in this period. Under the direction of Adolf Meyer as head of psychiatry, the university had opened its Phipps Clinic—the first full-service clinic in the United States, with an asylum and research wing together—in 1913. Meyer had already served a stint in New York as head of the Psychiatric Institute there and was engaged with Freud's notions of sexuality in infantile life. Perhaps as one of the famed French neurologists Jean-Martin Charcot's *other* students, he worried about Freud's total rejection of Charcotian biology. Johns Hopkins was also a hotbed of behaviorist activity, and the place from which John Watson launched his career. Under Meyer, Leo Kanner gained his first position at Johns Hopkins in 1928, and then became the director of the Children's Psychiatry Service, which was funded by the Macy Foundation and the Rockefeller Foundation, before their turn to cybernetic studies of the child. It was from there that the first schematic book on child psychiatry, aptly titled *Child Psychiatry*, was published in 1935 by Kanner.

In the 1930s, child psychoanalysis was coming into its own. It was initiated in the preceding decade, and now Anna Freud in Vienna and Melanie Klein in London began to formulate new theories of the child psyche, taking analyst and educator Helen Hug-Hellmuth's nascent practice of play therapy and Sigmund Freud's few direct writings on children (*Three Essays on Sexuality*, and his case about a young son of his student, "Little Hans") to the treatment of children in the consulting room. The first course in child analysis was offered in Vienna, taught by Anna Freud, and neurotic wealthy children from the United States were taken overseas to receive the treatment—their mothers accompanying them and sometimes receiving the training before coming home. Melanie Klein increasingly published across the 1920s and 1930s after qualifying as an analyst, and Anna Freud's *The Ego and the Mechanisms of Defense*, first published in German in 1936 (the English translation arrived a year later) marked the consolidation of ego psychology—which comprised mainstream psychoanalysis in the United States.

In the postwar era, as mothering in the United States turned away from scientific mothering and the highly regimented forms of oversight from doctors and scheduled care it required, the predominant model returned to libidinal or instinctive mothering, as inaugurated by Dr. Spock's 1946 imperative: "Trust yourself." Those instincts—and that form of unstudied yet practiced care—were understood to surround, control, shape, and impart messages to the child. (That mothers were still responsible for their families and yet unable to perform that very responsibility remained.)

In parallel, child psychiatry and psychoanalysis debuted on the world stage. The majority of leading European psychoanalysts, especially those offering major revisions (even conservative ones) to the theory of mother-child relations, were women and, indeed, often mothers (or in Anna Freud's case, a stepmother). These "maternalists" often disagreed with one another, sometimes brutally, but their theories carried weight: from Anna Freud's notion of acting "in the best interest of the child," which was used to restrict paternal custody rights after divorce in the United Kingdom for a generation, to Melanie Klein's descriptions of the mother-baby that gave rise to the theory of reparation. To turn to a maternal focus was a radical departure from Sigmund Freud's own interest. Despite popular conceptions, Freud had focused nearly exclusively on the father as site of neurosis. Now it was the mother's turn.

Each of the groups I have identified—psychiatrists, child development researchers, behavioralists, and psychoanalysts, and the many individuals associated with them—contributed these new theories of *mother-infant* relations, and with them, they invented new stories about what it was that mothers did with and for and to their children. They largely agreed mother was crucial, central, and determinate in child outcomes. How or why she was so, or what those outcomes might be, varied widely. Even just one field—psychoanalysis, as an example—was rife with disagreement. Since the 1920s, psychoanalysis had begun to fragment (let alone the earlier splits between Freud and his followers). "Freudian" itself could now mean several things; Freud had changed his mind so drastically about his own theory. Psychoanalysis—and particularly child psychoanalysis—was split in twain. There were those who thought everything was attributable to the psyche (Melanie Klein) and those who thought the fate of children was entirely environmental (Karen Horney). What remained between them was the idea of a mother—"real" or "psychical"—at the center of those arguments. They were joined by the above-mentioned behaviorists, psychologists, and pediatric psychiatrists, who also entered the fray with new theories of mothering. Almost always in the United States, these theories of mothering discerned and determined pathological outcomes in children.[18] Together they made mothering total. These inharmonious theories suddenly sang together. In James Clark Moloney's words, rather than for God and country, children were snarled in a constellation of "Mother, God, Super Ego."[19] On his understanding, they were one and the same, and the infant had a theophanic relationship to his mother—she was total, omnipotent, everywhere.

It was across this set of years, starting with the turn to a new focus on the mother in psychoanalysis in Europe and the emergence of psychological scientific mothering in the United States, I argue, that the human sciences began to reconfigure the mother as a medium of care. These social scientists made this pervasive argument from a new set of data presented by the war, and in response to a set of new social pressures—as materially, mothering became increasingly exclusive as a form of social reproduction (see chapter 1), especially for the therapized classes, and as mother was tasked with both caring for her child and protecting it from corrupting societal influences. Put another way, where under Freudian theory mother was the site of care and father came in to mediate society, mothers now were understood to largely play both roles, especially before children became school age.

On both sides of the Atlantic, new idioms, which I join together as a "media theory of mothering," tried to revise and precisely address how the mother was taken up in the mind of the infant. Often they were predictive and normative theories of the child psyche, or its development, and they were distinguished by a coinage that redescribed mother via recourse to the metaphoric and sometimes metonymic as well. Whether through émigré empirical psychoanalyst René Spitz's, British psychoanalyst British psychoanalyst D. W. Winnicott's, and American psychoanalyst Phyllis Greenacre's notions that the mother should be a total "environment," James Clark Moloney's understanding of the mother as an infant's "climate," attachment theorist John Bowlby's assertion of the mother as a "psychic organizer," as scene of "security" or its absence, Wilfred Bion's notion of mother as "container," or Adolf Meyer and his students' investigations into pathological mothering (eventually begetting the trope of the "refrigerator" mother), over and over mothering was revised to a function, a transmission, a device, an atmosphere, a milieu.[20] She enveloped the child, enclosed the child (Didier Anzieu);

she leaked the enigmatic (Jacques Lacan, Jean Laplanche); and embodied the popularization by Freud's student, Erik Erikson, of the term "maternal matrix," and psychoanalyst Margaret Mahler's use of the "infant-mother psychic matrix," which was used by many in this moment in both forms. Theories of development (of biological organisms, of children) have sat close to theories of the medium for centuries. Here, they joined one another fully.

The medium concept also has latent social meanings via its connection to the concept of milieu. Georges Canguilhem, the anti-fascist philosopher, called the milieu the "elemental medium thread" or a "perceptual envelope."[21] Christina Wessely writes on the milieu concept, which shares many features with the history of the medium concept: "The history of the milieu concept is substantial. Closely related to *ambiens* . . . the Latin idea of the *medium* came to consciousness through the natural sciences and theories of ether. Over the course of the eighteenth century, this became the French concept of the *milieu*. . . . Auguste Comte contributed substantially not only to the sociological but also to the biological contours of the milieu concept as the sense of an enclosing space."[22] Matrix is a synonym for the womb, yes, but its general dictionary definition, according to the *Oxford English Dictionary*, is "an environment or material in which something develops; a surrounding medium or structure."[23]

It is this definition that opens up connections between these competing theories, whose overarching aim is to redescribe mother as this total, conveying, imprinting, *medium* of care. *How* that took place, what good mothering looked like, the theory of why it was good, and what it might mean for a child to have this care withheld or offered was contested. While each put forward their own causal theory of mother-as-medium gone awry, that mother could be understood as medium is a paradigm clear across the board.

Rather than being kept in disparate clinics that only impacted those treated there and in person, these theories, as we will see, were domesticated frequently, in newspaper columns, in advice books, and on radio shows. Their authors also reached national prominence as advisors on national policy, delivering expert commentary and participating in both public programming and closed-door roundtables. Moreover, these strands—Yale and its child development emphasis; Johns Hopkins's program in child psychiatry;

US, British, and émigré psychoanalysis—may seem separate but they often combined in the United States, especially up and down the Eastern Seaboard, hooking up with Berkeley and the Bay Area at points.

The following are just some examples that will crisscross this book. The Viennese psychoanalyst Margaret Mahler, upon emigration to the United States, met with Dr. Spock and, inspired by his work with children, became a senior child analyst at New York Psychoanalytic Society and Institute. Spock also served as pediatrician to Gregory Bateson and Margaret Mead—thus bringing cybernetics and child-rearing into direct contact. Both in New York and in Baltimore, Adolf Meyer additionally trained and mentored a number of those who would go on to be central to this turn in the psy-ences, such as Phyllis Greenacre and Fredric Wertham, and he gave certain practitioners—such as Leo Kanner and Lauretta Bender (see chapters 3 and 5)—crucial institutional posts. (Many of the greatest proponents of mother-as-media theory were Jewish Viennese emigrees, as we shall see). Erik Erikson, Dr. Spock, James Clark Moloney, and David Levy (whom we will meet in chapter 3) were invited to the 1950 White House Conference on Children and Youth where they would *recommend* investing in television for children—nineteen years before *Sesame Street* debuted (see below and chapter 5). Bender and Fredrick Wertham were both trained by Meyer, and then overlapped again at Bellevue Hospital in different sectors—before facing off in front of Congress in 1954 on the question of the influence of *comic books* rather than mothers (see chapter 5). Many of Meyer's most famous students generated work central to this unwitting "mother-as-medium" paradigm. Meyer himself would be closely associated with maternal blame as a theory during and after his lifetime (see chapter 3).

Although these normative theories were often deeply punishing and prescriptive, feminist psychologists and psychoanalysts gave similar accounts of the primacy of mother-infant bonding, even as their social and political efforts often flowed in the other direction. It is important to note that what occurred in the United States in terms of the feminization of the psy-ences was quite different from what had occurred in Europe. The feminization of US psy-ences does not follow a neat path, and it does not map to typical accounts of deindustrialization. In the United States, too, the category of the psychoanalyst was delimited to the psychiatrist (in Europe, one could enter the field without a medical degree), and institutional psychoanalysis deployed the white supremacy and misogyny of medical education as a shield for its own practice.[24] As Martin Summers argues, in the wider psy-fields, "the white psyche was the norm," which had massive consequences for diagnostic racism within psychiatric psychoanalysis and beyond it.[25] Lay analysts, largely women, were far from content with this state of affairs and made their own paths, their own institutes, conferences, journals, and activities. But neither academics nor psychologists were welcome wholesale—if at all—in Freudian institutes accredited with the American Psychoanalytic Association (APsaA) and thus the International Psychoanalytic Association (IPA), nor were they swelling the ranks of medical programs. This deeply impacted the political and social tenor and uses of new theories of mothering, as well as treatments for mothers, who were often prescribed medication to help manage the anxiety that they were seen as passing onto their children.[26] As historian Ellen Herman details, psychology had, from the point of view of some feminist psychologists, named mother "public enemy number one," and it was time to "go look for the real enemy."[27]

In the United States in the 1970s, feminist psychologists, whom historian of maternity Sarah Knott in a forthcoming study refers to as "achievement feminists," focused on parity at work and in their professions, deeply revising theories of their (often male) teachers.[28] A litany of new terms and propositions emerged: Jessica Benjamin's "bonds of love"; Marion Milner's revision of the maternal matrix as the "primary matrix" in which the mutual mediation of infant-mother occurred;[29] Julia Kristeva's rewriting of the stages of development, especially in "chora"; Carol Gilligan's revisions and critiques of the Eriksonian maternal matrix; Daniel Stern's "maternal constellation" or "mutual connection"; Nancy Chodorow's elaboration of the mother-daughter relation; or the work of Esther Menaker, who argued that a mother's communications become an environment for a child.[30] A throughline emerges from these works, even as their doxa and idioms diverge: feminist theories of mothering not only continued to elevate the bond, like the maternalists' had, by making it exclusive, but also redescribed it as a view of *mutual shaping*, or in the words of the Boston Process Group, "reciprocal relations."[31] Something similar would go on to happen within media studies and science and technology studies; media would eventually be seen as both shaping and being shaped by their user.

Alongside these feminist reimaginings, psychologists and psychiatrists developed new theories of what toxic environments (as in pollutants) and hormonal differences in utero might do to children, and, as historian Angelica Clayton shows, they even figured trauma itself *as* toxicity.[32] Here the atmospheric media of mothering—popularized by Winnicott—became material and rendered bio-scientific what psychoanalysis had long proposed metaphorically: mothers serve as a hinge between the outside and inside, but also between bio and psyche, between human and nonhuman. They mediate and are the medium themselves.

While we now understand many of the theories described in this book—on both the psycho-medical and media sides—to be obsolete, drenched in racist and misogynist pseudo-science, or delivered to the public shrouded in those warped ideologies, linkages between mother and baby via the mother's mind and her environmental impacts on her fetus, infant, and child have been a persistent assumption in the history of medicine, no matter how much the mechanics of causation change. Mothers were understood to facilitate, to regulate, and to contain. Nondesirable outcomes for children were thus traced back to that very primary connection. Under this description of the psychosocial mother, mother became a screen for society because society supposedly originated with her, and could be mediated through her and controlled by her as she imparted it to her child, who then would go on to live in one. This was then reoffered and circulated as a description of mothering. Maternal purity, or a kind of maternal immediacy—the idea that we can communicate without mediation, without error—is an impossibility, a fantasy immediately used to discipline actual mothers. Diagnosing maternal mediacy was central to psychoanalysis, behaviorism, and the emergence of cybernetics, as well as in our own fraught understandings of mothering today. I argue across this book that "good" mothering was increasingly defined by the right qualities and quantities of presence and attention, by the right ratios of mediacy and immediacy. The problem, however, was that no one could agree on what those might be.

MEDIA AS MOTHER

What I am calling the mother-as-medium paradigm emerged in parallel with the conception of media as proxy mother. If it may be obvious to us now that media are called in to occupy, pacify, educate, entertain, and activate children (as well as watch and soothe them)—in sum, remediating the *functions* of "mother"—we can see the inverse process at work. Mother, broken up into her constitutive actions (especially in infancy, but not only), was assumed to be a series of functions, and the function was generally adduced to be the expression of care. At mid-century, psychologists added to these functions their understanding of mother as *substance*, of *conveyance*, of mediation. She was replaced or augmented with media, yet still charged with mediating the inside of the home and whatever came from outside—again, often media were the instruments of both choice and necessity for this process.[33]

Under the mother-as-medium conception advanced by social scientists at mid-century, the pathologies of one passed to the other, and vice versa. It may feel strange to think of a theory of media emerging from the psychological sciences, but as literary theorist John Guillory notes, until the middle of the twentieth century, the word and concept of "media" partially belonged to the tradition of psychology, where it was "most useful in constructing a picture of the mind in its relation to the world." The concerted emphasis of many branches of psychology in this moment was the picture in the infant's mind of the mother (and, as we shall see, also media). It is from psychology that we have a subsection of the definition of mediation: "The interposition of stages or process between stimulus and result, or intention and realization."[34] The mother was the stimulus—even in non-behavioralist accounts—and the baby was the result. This logic could be used to signify, dissect, understand, and patrol the role of the mother and her labor both reproductively and productively. As we have seen, psychologists simultaneously reduced mothering to a fungible function and expanded her import to a total, irreplaceable environment. Once a set of fungible functions, mothering could be automated, could be reallocated to nonhuman others. As we saw with Bradbury's story, this was a double bind.

Where researchers like psychologist Harry Harlow railed against "wire mother" proxies, extrapolating from his research on monkeys, psychologists were starting to condemn *wired* mother proxies for human children. As with the earliest notions of mother as environment, mediated mothering—using media as part of child-rearing—has been part of diagnostic regimes since at least the 1920s, and it has even earlier antecedents in the use of optical toys.[35] Media's effects on children were of great interest not just to pediatricians but to proto-media theorists in this period—to media studies as a field—and discourses surrounding what we might now call *toxic* media began to crystallize, even if they did not have this name.

The remediation of mothering, especially when it is the mother who remediates herself, was understood with great frequency, although not ubiquitously, as the worst of psychological and developmental crimes by both psychologists and media researchers.[36] In Bradbury's "The Veldt," after all, it is a psychoanalyst who is called in to state the obvious: the machine must be turned off. Alongside the psychologists and psychoanalysts mentioned above, those studying media have often addressed US mothers with a similar message. The use of media with children, or mediated mothering, has long been understood through the psy-fields and their attendant diagnostic categories, whether in representation or reality. It is no surprise that media theorists, too, had a great deal to say about what was occurring in the mid- to late twentieth century between mother, child, and media. Again, this was no coincidence. As John Guillory defines remediation, "Remediation makes the medium as such visible."[37] It makes sense that the social process of remediating mother also put mother into relief—that critiques of technologized and mechanical versions of care also cast aspersion on human forms. The reverse also holds: the pathologies associated with mother attached, in turn, to her remediations.

We can retell the history of media research in the United States as investigating this very intersection of mother and media. Starting with the earliest media researchers in the United States studying the impact of cinema on infant sleep in the Payne Fund Studies of the 1920s, a new and major concern for psychologists and sociologists lay beyond the family, looking to other sites of "interference" or influence. This meant that other scenes of instruction, and therefore *other* media (the cinema and the TV, yes, but also the devices used directly for care, such as the crib or the baby monitor) were of interest to those engaged in audience research. Media were, in these early studies, already configured as impinging upon the paradoxically primary, naturalized, and central *human* mother medium, even as she too was disparaged.

Mothers, too, were concerned about how to manage these new interlopers. When the staff at the Child Study Association of America collected the key questions to which mothers wanted answers about their children across the 1930s, mothers wrote with great frequency about what to do about "outside influences, such as that of newspapers, school, movies, friends."[38] Media researchers had the same questions, formulated in nearly identical terms across a number of different schools, including the earliest media psychologists and media effects researchers (like the US-based members of the Frankfurt school), such as Paul F. Lazarsfeld and Theodor Adorno at Princeton.[39]

In this moment, psychoanalysts began to think of movies as contributing to the fears and misbehaviors of children, "especially those in public school," who were more likely to follow "id impulses"; these findings, in different terms, were echoed by sociologists who performed the first systematic media research.[40] In the 1940s, movies were seen as an influence on a range of less conscious behaviors as well—from tics to other compulsive repetitions.[41] When cinema normalized as *less* of a threat, by the 1950s, psychoanalysts became interested in cinema as apparatus, forming a backbone of cinema studies. Comic books and eventually television joined cinema in what would later be termed a *media ecology*.[42] When psychoanalyst Fredric Wertham and psychiatrist Lauretta Bender finally testified before Congress, as I mentioned above, they were called as experts in children, yes, but as experts of children's use of comic books rather than on psychosocial disorder (in the case of Wertham) or schizophrenia and autism (in the case of Bender). Their pay-to-play expertise concerned both mother-as-medium and media-as-mother. Similarly, just a few years prior, when Dr. Spock, Erikson, Levy, and others gathered at the White House Conference on Youth and Childhood, not a single media psychologist was called upon. Instead, the conference favored those in pediatrics and child development. One of their signal recommendations: increase the usage of TV as a medium of care and education in the home.

By the 1950s, as Lynn Spigel shows, when television was rife with contradictory meanings, so was mothering. For those who were anti-media, mother was the answer. For those who thought there was not enough of mother, or the wrong kind, media were often acceptable. And for some social scientists, there was not one, but two degraded, cold, mechanical mothers in the home. As we shall see, there was a serious correlation: those who believed in influence-theories of mothering often believed in influence-theories of media (even if those influences were *positive*).

While ideology, affect, and widespread pathology were all central sites of panic surrounding the contagious effects of the wrong kind of stimulus and mediation, media themselves and their role in child-rearing were subject to the same debates. By the 1970s, we can say a discursive war was being waged in the American household: Mother v. Media. Both were suspect, both were indispensable, but remediation became configured as worse than the mother-medium. The worry, of course, was that media had won—and mother was elsewhere (working, divorced). These two strands of psychology—one focused on mothers, one focused on media—were not so separate. They shared a language and a set of taxonomies. In the inter- and postwar debates on the role of mothering in producing delinquency and other "undesirable" outcomes, media were a central focus, too. By the 1990s, when the term "screen time" was coined and applied retroactively to children of the 1950s, it was under the sign of Tom Engelhardt's 1991 screed entitled "The Primal Screen": television had literally replaced mother (and father) as a primal influence.[43] All neurosis was thus, given the exaggerated Freudian punning, supposed to come from television. Media panics did not displace the panics about, and surveillance of, mother, so much as add new ways to blame her for her mediating impact. The only normative and prescribed antidote: mother needed to mother purely, and more often.

The history of media studies starts long before the field consolidated. Traced back across the twentieth century, what we call media studies now has its roots in many fields that turned to the study of media technologies, the media (i.e., the content it carried), and mediation. The Chicago School and early media psychologists were thinking about the problem of influence as originating in *the media*, its content, as well as its form, whereas the early media theorists turned toward thinking about the vehicle, or the medium—its conveyances. This eventually becomes a study of "media effects" or the influence of media on individuals and collectives, and a study of "media ecologies," pioneered by Marshall McLuhan. McLuhan borrowed heavily from multiple forms of psychology and psychoanalysis for his notions of figure/ground (i.e., "the medium is the message"), and in the 1960s, Neil Postman examined the interaction between human subjects and communication, media, and technology. Again, rather than defining them as being separate, social science deeply informed these early turns to media theory, and the terms of the medium, as well as *environment*, passed between them. For Postman, "if in biology a 'medium' is something in which a bacterial culture grows (as in a Petri dish), in media ecology, the medium is 'a technology within which a [human] culture grows.'"[44] This logic of medium and culture can easily apply to a mother, in whom and with whom it is presumed a child grows. In the 1960s, as D. W. Winnicott became synonymous with earlier theories of the maternal environment, *media* ecologies were now structured similarly.

Across the period when these new theories of media emerge in the University, new attention was being placed in the home and in the clinic not just on how to get the television to bring the family *together*, but on how it might supplement it, become an individual part of it. By the time *Sesame Street* debuted—after years of research, in part conducted by psychologists—child development specialists, technologists, educators, critics, and producers all had something to say about media—and the media where mother was not. As media studies began to distinguish itself as a discipline—a slow work of melding communication studies, film studies, English, and comparative literature—media theorists commented on both mothers and children where they might find them. Therefore, we have a long-standing record of both media theorists *and* psychologists—often the same ones investigating mother-child relations—commenting with no less depth and detail on the ubiquity of techno-media in the American home.

Focused largely on television in the 1970s, but also on the emerging cultural turn to video games in the 1980s, media *ecologists* looked at a different metaphorical ecosystem, and tried to diagnose its misproduction of subjects—or the production of aberrant ones—like a bad mother produced her bad child. To make its paradigm stick, the media-as-mother concept exploited the crisis of childcare, in part produced by the mother as medium paradigm. Reproductive labor and the maternal environment slowly became the exemplary sites of attention for psychoanalysis, psychology, and psychiatry—and how mother extended her literal minutes devoted to care by augmenting or supplementing her care regimes with media. As Lynn Spigel shows in her groundbreaking *Make Room for TV*, bad media were often used to metaphorize the state of the family; "bad signal" on the set became "communication breakdown" in the family.[45] I would add to this account that it also went the other way. Rather than only metaphorically, eventually, the diagnoses that once attached to bad mothers (autism, say) now attach to bad media. The panics over mother-as-medium are now proper to media-as-mother. Diagnosis—no matter its direction—serves as one regulatory mechanism on mothering, especially mothering considered in terms of regimes of attention and devoted presence, or assessed for the purity of the environment a mother offers. She is judged on the ratio of her signal of care to the noise of herself.

MAKING FAMILY

Where mothers are, in the modern United States, pathology is rarely far behind. As "The Veldt" so evocatively suggested, thinking mother and her remediation together was part of a conservative and complex postwar moment. Following Erving Goffman, we can think of the family as the first total institution.[46] It is from the family that we are socialized. Therefore, the family is also understood to be that structure that mediates the social. Starting in the twentieth century, across sociology, medicine, and psychology (including psychoanalysis), theories of the family all seemed to agree that the stronger the cell walls of the nucleus were—the more the family held strong within itself—the better. Conversely, the more permeable the family was to mediation from the social, the more catastrophic for the family, and especially the child within. It was socially understood that the wrong kind of mother—and the wrong kind of media—could corrupt not just individual children, but children as a class, as a generation. In the postwar moment and Cold War era, intense research was conducted on the democratic and authoritarian families and their communication structures, how they mediated the outside and inside, took the psy-ences to the scale of the nation. The problem with mother-as-medium was first perceived in experiments on individual children, but it soon was understood to play out on the national stage.[47] As historians Elaine Tyler May and Fred Turner have shown, mothers were figured as essential to the protection of democracy at home, and the family was configured as an anti-Fascist democratic technology.[48] When the family, and the mother most centrally, mediated the social, mothers were also understood to produce its degradation.

Feminists, womanists, and historians have long reinterpreted motherhood and care, detaching the concepts from their naturalized but false location in the single figure of a birth giver. In its usage, mother has variously been a familiar greeting to elder women and the term for someone who gives birth metaphorically or literally to *anything*. This site of personification has been around for almost 500 years (Mother Earth, mother tongue). We can mother and be mothered irrespective of particular relations to a child. The maternal does not even have to be gendered: the maternal can have no gender.[49] As Lisa Baraitser writes, "When I use the term 'the maternal' I mean it in its widest sense—to include motherhood as an embodied and embedded relational and material practice (the very literal daily, and often arduous labour of raising children by anyone who identifies with the term 'mother'), through to its figural, symbolic and representational forms."[50]

Historically, the maternal—irrespective of the labor performed—*has* been gendered, raced, and classed—and violently so. This was one of the first and longest standing insights of American women's history and Black feminism. As Hortense Spillers reminds us, motherhood is a reified status granted only to middle- and upper-class white women, and Alexis Pauline Gumbs then turns us toward thinking about mother*ing*.[51] Or, as Sarah Knott puts it, *mother* is a verb, a set of actions and tasks such as nursing, sleeping, rocking (each of these actions, as we shall see, has been remediated).⁵² It is not so simple to say that mothering has been unidirectionally attached to the feminine or coincidental to femaleness—itself a violence and an exclusion. This book traces the aftermath of the Cult of True Womanhood and the Cult of Domesticity or its revision, Real Womanhood, from the antebellum period through the start of the Progressive era. The primacy of the mother as defined under these social logics is always via the occlusion of others. The role of the father (especially under the sign of panics that he has disappeared), the possibilities of other kinship arrangements, and many of these same mothers and carers (enslaved mothers, waged caregivers, and others) are negated under these constructions.

Two particular trajectories are especially important to the account told here. Each has its antecedents in the antebellum and postbellum period and each continues to shape daily life in the United States in our present. The first is that of natal alienation in the long wake of the plantation, wherein a mother might have no rights to her child such that she could lose her child, or the child could lose her.⁵³ Natal alienation is the subtending compact of American family life and the backbone of family policy in what is now the United States. Forced relocations of members of Indigenous kin began under Spanish colonization, and then continued through the widespread use of boarding schools and the "60s Scoop." Routine family separations have been and remain a major feature of carceral control and the punitive discipline of Black families from Reconstruction all the way through to our contemporary moment, and Black children are many times more likely to be reported to Child Protective Services and taken away from their homes.⁵⁴ At the US-Mexico border, in ICE detention, family separation is the norm. For the last 100 years or so, psychologists have been ready to frame and revise iterations of these policies. Thus, the second arc moves in the opposite direction: mother is commanded to be everything for the child (and if she fails, sometimes it is under the threat of forms of separation from her child, as we shall see). A terrible scapegoat and often a victim of psychosocial displacement, mother is configured as responsible for the effects of these policies. Accordingly, the mother-as-medium frame contains both these movements. As we will see in chapter 4, maternals are managed and regulated *as if* they are media, including when their contact with their children is reduced or restricted to mediated forms.

As a site of purportedly unmediated purity, childcare is part and parcel of the fantasy of the largely white, middle-class housewife—or, as Angela Davis terms it, a "partial reality."[55] Davis uses this term to describe the historical occlusion of any woman who has the double burden of not being "just a housewife but having always done their housework," as well as those women who were unwaged via the subjugation of enslavement.[56] As Stephanie Coontz has written, too, this fantasy, despite its pervasiveness, does not reflect the economic reality of even the group it supposedly addresses, not even in the time or place from which it came.[57] This restrictive and privileged fantasy centers on preserving dependency of mother and child on one another, and accordingly, it forecloses the possibilities, both real and imagined, where mother and child might be disarticulated from that dependency. Here, the partial reality of unmediated childcare joins the long-standing partial reality of housewifery, and it demands that we question the particular form of this most "primary" of relations: Which child, what childcare, and which provider?

Now, in our present, parent-consumers spend $10 billion annually on provisioning their homes with technologies for child-rearing, from bottles to baby monitors—a quarter of what is being called the "new mommy market" or "new mom economy," cute names for the $46.5 billion controlled by the millennials who are becoming parents at the rate of one million new consumers per year.[58] But consumerism and parenting technology are not new. Devices that did the work of standing in for mother became ever more prevalent in the first decades of the twentieth century. Many of the most crucial, ubiquitous devices on which we now rely—high chairs, cribs, strollers—have been in use for more than 100 years. High-tech "innovation" on these items commenced at the beginning of the twentieth century, and parents have turned, since the interwar period, with increased frequency to a range of media and domestic technologies meant to automate care, entertainment, and containment to accomplish the never-ending tasks of parenting. Media technology in the home and for childminding provide a spectacular test case for understanding the double bind of parenting as it is articulated through mothering—that is, that the labor of making family is unending and yet must increasingly be accomplished without human help from outside the family. Even as those tasks change, the ideologies surrounding them shift, and the economic conditions inside and outside the house volatilize, mediated mothering remains constant—as does the notion that mother is a medium herself.

Remediated forms of childcare designed to ameliorate parental labor and fear produce in turn new forms of parental labor and fear—of surveillance, of medicalization, of separation. This cycle has been central to the lived experience of many American parents since the interwar period. Childcare is, in the US context, undersupported and fractured and has been across its long history—even as the fault lines have changed. Some people are just coming to that fact after years of pandemic; many have campaigned for the very possibility of sufficient care for much longer. As feminist, particularly Black feminist and Marxist feminist, scholars have argued for the last fifty years or more, the crisis of childcare is a crisis of labor, a long-standing one, but it only gets termed as such when it comes for the middle class (as it did in the interwar and World War II contexts, as I detail in the next chapter, and as it has in ours). That labor—which the term "parenting" conveniently obscures—was reconfigured by the psy-ences across the twentieth century as attention, presence, and environment. Sometimes the work of parenting is to use presence to mortgage attention: send it elsewhere while remaining with a child (as in telework; see chapter 6). Sometimes it's the converse: the maternal is already elsewhere (at work), but still paying attention remotely via a digital feed.

An oft-touted statistic is that parents of any gender now spend more time with their children than their counterparts did in the 1950s, 1960s, or 1970s. Sociologist Sharon Hays aptly calls this "intensive mothering": no matter how overextended, each individual mother (or parent) is held up against the cultural ideal of 24/7 mothering—the always-on mother. As we shall see, this is understood to be increasingly possible, even as it is not feasible, because of the use of media itself, and media have added to the work of mothering: controlling screen time is now understood as part of the job.[59]

For all the social and economic change (and change in media) from the economic miracle of the postwar era to its end, what makes a mother remains under contest. Despite mass change to the family form in the period under study, US culture was obsessed with "motherhood." Then as now, this is a culture that delimits what a mother should be, look like, and do—whether such a person is possible or not. In this book, there may be great resonances for contemporary readers facing panics about what constitutes proper mothering today. In the debates surrounding technologized childhoods, let alone in relation to the pandemic and to distance education, we can still read the paradoxical idealization and criminalization of mothering. These debates and panics are still filled with the echoes of the human sciences that insisted mother had, on my understanding, become a medium. The same holds for how we think about media use now—by adults and children. The social panics of the comic book or television may strike us as silly or quaint. Even as social panics unfold around us, we might write them off as just that, waving them away. But panics are crucial sociocultural symptoms, and studying them will always tell us a story.

Mother Media tries to understand how the social panic about bad mothering informed our understandings of media, and vice versa. As I argue, these theories take on new meaning just when media were used socially to extend mothers. *Mother Media* picks up on dispersed thoughts in critical theory, some complete and others enigmatic or gestural, about how media became mother-prostheses, as Marshall McLuhan might say, both culturally and socially. As Fredric Jameson has pointed out, the rhyme of reproduction (of an image, say) and reproduction (in the sense of childmaking and rearing) cannot go unremarked; Donna Haraway took this one home in her cyborg manifesto, written avant la lettre. Klaus Theweleit pointed to this psychical, rather than technological, collapse of media and mothering in his book *Object-Choice (All You Need Is Love)*, arguing that filmmakers and other artists tend to fall in love with *medial women* (the women who edit, type, and otherwise work with their media).[60] If all objects are objects refound, Klaus Theweleit's auteurs were in some way refinding their medial mothers (though Theweleit is also invested in thinking about siblings refound). Almost at the same exact moment, media theorist Friedrich Kittler proposed in his book *Dichter/Mutter/Kind*, that mothering specifically and the emergence of the bourgeois family in the nineteenth century itself contained his technical a priori.[61] Here, I offer a historical account, rather than a theoretical one, of mother-as-medium and media-as-mother's co-emergence in the United States.

This book also provides us with a narrative of how the social panic around media first attached to children—and then remained in their domain, and that of their parents, for nearly a hundred years. In our era of the intimate smart phone, this story has become, as television scholar and critic Phillip Maciak has written, "everyone's problem."[62] Maciak observes that the way we think about screens colloquially now, and our use—always implied overuse—of media, is formulated on the grounds of the very children studied by the actors of this book. They are, in this way, a prehistory of ourselves.

MOTHER'S LITTLE HELPERS

Melinda Cooper writes, "The history of family is one of perpetual crisis."[63] As she shows, that crisis is very different depending on the family in question. These iterations pervade the twentieth century and extend into our present, even as fears about what might replace mother shift from the nanny and nursemaid—equally as worrisome to parents at the turn of the last century—to the smart home, and devices and screens. The remediation of care proceeds along the double valence of that term, as both the act of remedying and the act of reformulating the old in the new on the grounds of media. Whether paid for, given, or automated, care is deeply interlinked with concerns about control over the object of care and the caring subject alike; it both redresses *and* produces fears of intrusion and the corruption of children. Most frequently, it returns us to a fantastical notion of the purity of mother and child operating as a single unit, a dyad that must be protected from incursion at all costs.

From the nursery to the prison, from the clinic to the commune, *Mother Media* examines twentieth-century pediatric, psychological, educational, industrial, and economic norms around mediated mothering and technologized parenting, charting the crisis of family mediation. The first chapter looks at attempts to remediate nursemaid and mother via speculative technologies, model and futural architectures, and screen media. The second turns to a history of soporific media, labor, and child-rearing practices that worked to alleviate the demands on mother, while also conforming infants and small children to the rhythms of capital from the Second Industrial Revolution forward. The third chapter tracks the shared metaphorization between media studies and psychiatry at mid-century in developing new taxonomies of "bad" mothering, which paved the way for these new pathologies to attach to media. The fourth chapter extends this ground by looking at how notions of "environment" and "security" were studied in a US prison, where mothers and babies lived under punishing, controlled (and thus eminently observable) conditions. The final two chapters consider how these notions of maternal environment and maternal pathology—and the new understandings of maternal attention, presence, and absence—moved from mediated mothering to mothering by media, with a focus on the psycho-social panics surrounding the inclusion of technology in childcare.

Let's go look at the nursery.

I THE MEDIA OF MOTHERING

1 THE FIRST TECHNOLOGY IS HUMAN

On March 1, 1932, Charles Lindbergh Jr., the twenty-month-old son of aviator Charles Lindbergh, was kidnapped from his crib.[1] The night had started like any other. Charles Jr. had been put to bed; his parents were home, with a full staff attending to the family, including a butler and the baby's nanny, Betty Gow. Gow, in her routine check on the baby before bed, discovered the empty crib and a ransom note while Anne Lindbergh was bathing, and reportedly burst into the bedroom exclaiming, "Anne—they've stolen our baby!"[2]

Immediately, a wide-scale search began to find both the baby and his abductor. Gow tragically became the first real suspect as the Lindbergh family and the police turned on those closest to them, believing that the crime must have originated inside the house. When there was no longer any baby, there was also no longer any "our." Overnight, the *New York Times* picked up the story, with a headline that simply read: "Woman Believed Involved."[3] Minute-to-minute updates were broadcast on the radio; as Paula Fass argues, publicity is always part of the project of recovering the kidnapped, but here the media "completely overshadow[ed] the gentle emotions of those most intimately involved"—the parents. The media became another mother, and was both content and form of grief.

Charles Jr.'s other mothers were central to the investigation. Gow, as both care provider and the person who discovered that the child was missing, was implicated and questioned for days before she left the country for Glasgow, never to return.[4] Violet Sharp, a woman working in the household as a servant, was questioned so extensively and was under such intense suspicion that she ended up taking her own life by drinking poison.[5] She was cleared via alibi the very next day.[6] Family friends were interrogated for any possible connection to the disappearance. The nursery was dusted for fingerprints, as was the first of several ransom letters. No leads emerged.

What little physical evidence remained was destroyed: the Lindbergh estate, Highfields, in New Jersey, became a site for spectators, who trampled the grounds, trying to catch a glimpse of the grieving family.[7] Lindbergh Sr. was convinced mobsters were somehow involved and began a program of outreach to the likes of Al Capone and other major criminals, to no avail.[8] On May 12, the toddler's body was found by a passing trucker not far from the Lindbergh estate, haphazardly disposed of and mutilated.[9] News of the investigation reached President Hoover and the proto-FBI was spurred into action; every possible resource was put at the disposal of the Lindbergh family.[10] Eventually, the investigators gave up on their mob leads and followed the money: a series of bills tied to the gold certificates used to pay the ransom had been spent all along a single subway line in New York City, leading to the German enclave in Yorkville. After one such certificate was turned in at a bank, the police located Bruno Hauptmann—a German immigrant—whose apartment turned up other key evidence in the case (most damningly, a notebook with sketches of the outside of the Lindbergh home and a prototype of the ladder used to reach the baby's window).[11] It was rumored that he must have had help, and suspicion fell back on those employed as domestic servants in the home. Hauptmann alone was convicted of extortion, kidnapping, and murder.[12] In 1935, he stood trial for five days as crowds lined up to glimpse the family and its assailant. Hauptmann was found guilty and subsequently executed via "Old Smokey," the "affectionate" name given to the state of New Jersey's electric chair.[13]

The details of this case—one of the most famous American kidnappings—have been rehearsed and rehashed in the ensuing near-century. Speculation around the perpetrator has been extensive. Charles Lindbergh Jr. was not a representative American infant, except in the way that, especially in their tragic loss, rich white children are often made to stand for all children. The private family tragedy turned into both international sensation and national policy: each update in the case was tracked in the press, and new laws were passed immediately to protect future Lindbergh babies, making kidnapping across state lines a federal offense. The case persists in the American imagination not only because of the glamorous family in question and the noir aura of Great Depression crime, but also because of the nature of the crime and its location: the abduction of a baby from his place of assumed security—the nursery, with nanny nearby.

The Lindbergh kidnapping touched a nerve because it played into fears being stoked among far more ordinary families. The Lindbergh family was, by any measure, atypical. Charles Lindbergh Jr. not only was born to immense economic privilege and a father with a penchant for the limelight, but also, because of his wealth, lived in that rare house that held a family and its staff. By 1935, when he went missing, this was unusual: middle-class homes that had once employed staff, especially for overnight care, now only employed daytime help, usually for cleaning, if at all.

Nonetheless, the tensions surrounding family life in the 1930s are legible in this scene, and there is a historical and theoretical framework that subtends and explains the national fascination with the crime. As we shall see, the question of how the Lindbergh baby went missing was caught up in shifting meanings around mothering and maternal purity, proxy mothers and paramothering, and raced and classed "impressions" from proxy mother to infant. The results of the Lindbergh baby kidnapping—one of which was the introduction of automation of the already fungible nanny figure—were therefore never causal but instead part of this sea change already underway.

That the Lindbergh kidnapping was read as so spectacular yet so threatening, that it remains entrenched in the imaginations of white families, signals to us that it might be a screen, a cover for the set of family separation policies that subtended the reproduction of white family life (see the introduction to this book). Paradoxically, the possibility of the insider acting as violent conduit for the outside—here as a kidnapper—only affirmed the nascent conception of "the nuclear family complex." The nuclear family as a conception had only first appeared in England in 1924 in Bronisław Malinowski's "Psycho-Analysis and Anthropology." As historian Brian

Connolly shows, this now natural assumption of how the modern family might be shaped and function—with huge psychosocial ramifications, from the acceptance of the Oedipus complex to the family wage—was first born of a misreading of Freud's "nuclear complex of neurosis."[14] This simple misreading gave rise in turn to a new complex, which was immediately depathologized, and then took hold in the American imagination as, in historian Stephanie Coontz' terms, "the way things were"—and should always be—despite massive changes to family structures across the middle of the twentieth century. The notion of the nuclear family only further ratified the idea of the nuclear, the inside, a cell, in psychoanalyst Wilhelm Reich's terms, with strong walls. Families were told—by pediatricians, women's magazines, and psychologists—that good families relied only on themselves. And now the Lindbergh baby kidnapping, with its mass media attention, served as the ur-example of why: or else.

The exclusion of waged caregiving was not total; in practice, families increasingly patrolled their boundaries, demarcating the inside and outside. Those who, by the very nature of their contracted employment, supposedly breached the perimeter—like the nurse or nanny—were now suspect. That growing suspicion coincided with the rise of domestic consumerism and time-saving devices. While the middle-class family traditionally did not reproduce itself alone, now, after the great decline of domestic servants in the interwar period, biological ties and shared living space would matter evermore.[15] The late Progressive era also solidified child welfare, child psychology, and pediatrics; each used its bolstered authority to question how mother should care for her child, and whether anyone or anything else could. Despite much disagreement on other aspects, there was unanimity around one thing: mother should mother alone.

To affirm the nuclear family, for mother to mother alone, mother and nurse had to be further disambiguated from one another. Put simply, the nurse could no longer serve as a proxy for mother. In this chapter I argue that media were deployed to do the cultural work of standing in for the nurse, who was no longer understood to be an acceptable stand-in for mother. I argue that these technologies, often presented as futural, were nostalgic for the long era of the controllable proxy of the nanny. Joining nostalgia for this kind of waged human helper (minus her supposed contagion) was the "dream" of a fungible, self-reproducing labor force replaced by automata, which has its antecedents in the nightmare of the plantation and the workday in the factory.

Nostalgia, itself depathologized through the postwar era, originates in the early eighteenth century and its sudden social and economic change (alongside the concept of "generation" and generational change). It was, not surprisingly, a mode of feeling that was found in two groups most keenly: servants, away from home, and soldiers. Crucially, it was not mothers but the technologies themselves—and their designers—that were nostalgic. These technologies gesture at removing additional waged workers from middle-class and upper-class homes, ostensibly to recenter mother and make whole and pure the nuclear family. This purifying of the nuclear was coincidentally the actual introduction of a new "contagion"—media and devicing. As we shall see, the impossible logic of purity always leads to mediation panic (around mother or human proxy or tech proxy).

The goal in remediating the servant class (or the enslaved domestic worker) was therefore not to make a set of fungible automata that stand in for *mother*—which implies the exclusive class of women whose childrearing was supported traditionally by these modes. Instead, her other human helpers were to be remediated and made, if not into fully functioning machines, then into objects of surveillant control. As Neda Atanasoski and Kalindi Vora write, "new technologies have historically designated what kinds of labor are considered replaceable and reproducible versus what kinds of creative capacities remain vested with privileged populations and spaces of existence."[16] The nanny was both figured as a technology and designated for replacement. Atanasoski and Vora call these technologies "surrogate humans"[17]—a name that rhymes with the demand on the nanny to stand in for the mother without replacing her. That she was already a stand-in made the task only more complex.

Remediation is so much more than just replication and duplication—which might otherwise be how we think about these devices designed to replace the arms, or solid body, or rocking capacities (see the next chapter) of the nurse. While automating these functions was crucial, so was representing the past of the nanny's unobjectionable service, her taken-for-granted labor, her silent tact, her persistence. I will argue that extra-familial care poses itself as the midpoint between mother-as-medium and media-as-mother. Once the maternal is fungible, the passage from human to nonhuman proxies does not express a new paradigm but only new terms within the same framework. Nanny was already doing the work of standing in for mother and was also already conceived of as a medium (if there was maternal impression, so too was there wet-nurse impression).

In their interwar and mid-century iterations, we should think of mothering media as remediating the nanny/nurse, not remediating the mother. Remediation can be said to carry its full meanings here—as remedy and as transformation of older technologies (here, human, and for care) into newer ones. Once, the nanny was a perfectly serviceable proxy mother, indeed a usual and unremarked feature of a social hierarchy that despised labor. Once, she was an appropriate specialist in drudgery and the mundane. From at least as early as the Victorian period, understandings of mothering as "nonwork" or as a "labor of love" repressed recognition of the available, paid human labor of others (other kin and other childcare workers) as part of stratified reproduction,[18] paid attention, and care.[19]

This shifted rapidly in the first decades of the twentieth century. Just as the impulse to render the nanny functionally equivalent to a machine emerged, the nanny panic, which circulated in women's magazines and eventually informed pediatric advice, also began to function as screen for the inclusion of surveillance technology in daily domestic life—trained on the child and capturing the domestic worker in its wake. To solve the problem and panic of the nanny she must be turned to machine or else watched by one (in that slippage where machines are assumed to be neutral). Management of the nanny then happens remotely, whether that nanny is a human or a robot and thus completely obedient and not an object of maternal envy.

To understand why one would want to automate the nanny and her associated tasks—the most demanding of child-rearing—requires understanding how the figures of the wet nurse, nursemaid, and nanny became indispensable in many bourgeois family relationships long before Charlie Lindbergh Jr. went missing that night in 1933.

NURSE, GOVERNESS, NANNY, HELP

Paid, one-to-one caregiving for children is far from the major story at any point in US history, especially for working mothers. In the nineteenth century (and to our present) there were a staggering number of arrangements for care that allowed mothers to work and have children simultaneously. As historian Sonya Michel writes, "At the end of the nineteenth century, then, American child care had come to consist of a range of formal and informal provisions that were generally associated with the poor, minorities, and immigrants and were stigmatized as charitable and custodial."[20] Beyond arrangements of taking children to paid childcare centers—whether in department stores, one's own apartment building, the back of a local house—or to family members, usually elders, or having siblings co-raise one another, mothers have long relied on technologies to keep a child with its mother but to assist in care, freeing maternal hands for other work, remunerated or not. Michel movingly assembles a litany of the forms of care available before twentieth-century hallmarks of child welfare reform:

> Native Americans strapped newborns to cradle boards or carried them in woven slings; Colonial women placed small children in standing stools or go-gins to prevent them from falling into the fireplace. Pioneers on the Midwestern plains laid infants in wooden boxes fastened to the beams of their plows. Southern dirt farmers tethered their runabouts to pegs driven into the soil at the edge of their fields. . . . Migrant laborers shaded infants in baby tents set in the midst of beet fields. . . . Mothers have left children alone in cradles and cribs, and have locked them in tenement flats and cars parked in factory lots. . . . Mothers have left babies dozing in carriages parked outside movie palaces.[21]

When we now think about childcare, there is an implication that mother is dislocated from her child, that instead of the mother caring for the baby, a changing of hands occurs, and someone else comes in to do the work of helping the baby continue to live (if not thrive). Mediated childcare—whether by low-level or high technologies or, as we will see in this chapter, using media *as* childcare to change aural and spatial limits (or via its absorptive qualities, as we will see in chapter 5)—disrupts this assumption. Mother may be in the presence of her child, but with her attention directed elsewhere. As Michel deftly shows, there is a long history, in every class, in every US context, of using devices such that mother can labor and care at once (we will see how this crystallizes in telework; see chapter 6).

If low-level technologies have often been used for working-class and middle-class mother, the story of automating child-rearing is, centrally, one of expelling external, paid, and non-kin help (or "delegated care") from the mother-child dyad while preserving that aid in a new form. Put another way, the story of automated child-rearing is inseparable from the disappearance of dispersed mothering among the wealthiest of mothers, those who, before the 1970s, were largely not working beyond the home (for more on these technologies, see chapters 2 and 6).[22] Maternal help and its remediation paradoxically indicate the limits of the primary maternal—itself the central fiction of the nuclear family—while consolidating the supposed purity of that limited position.

Across this period, how psychiatrists thought a mother should love was open to a series of revisions. Often against and within an increasingly mechanized domestic environment, a mother was told that she must love more "naturally"; her maternal labor must be more intensive and attentive, total and unflagging. The reliance on the nanny as a prosthetic extension or as a replacement altogether for maternal labor—and subsequently, the technologies that remediate the nanny—begot a logic of pure mothering, exclusive mothering, banishing even nonhuman agents to the background. Elements of this dictum appeared for the working class as well, with the appearance of standardized daycare centers at the turn of the century, as well as the elaboration of the Children's Welfare Bureau, which, in part, aimed to control mothers by providing economic provisioning for them, so that they could leave the workforce—and yet, at the start of the Great Depression, fewer than 1,200 centers for early childhood care—across a range of forms—existed in the entire country.[23] These other forms of care mentioned above—intra- and inter-family care, waged care, and technology-assisted care—were largely the norm.

It is a strange and particular twentieth-century fantasy that exclusive mothering is the primary and usual way care of children might proceed. This fantasy, crucial to the ground upon which the mother-as-medium theory would appear, shaped the twentieth-century archives of the figure of the nurse and nanny, posing challenges to how we tell this history. The nanny is often pushed to the edge of the frame, just out of sight, spectrally present in the archive of daily life under capital and in the writings of psychiatrists charged with middle- and upper-class care in the twentieth century. This is one inflection of Anne Boyer's blunt definition of care: "the suffering called gender named by capital as love."[24]

To neglect histories of paid caregiving in thinking about childcare is a grave error. We lose much parenting history when we do not think of maternal labor as multiform. Not only does it aid and abet maternal ideals that harm almost all—not just mothers—but it also makes it impossible to consider working mothers who care for children in addition to their own, from the plantation through domestic service to the nursery school. "Mother love" is readily mobilized in more sentimental accounts of historical mothering to obscure maternal labor.

A figuration in which love and work collapse, where reproductive and productive labors collide, the nanny posed challenges deemed too significant to the emergent theories, and thus as a figure she was disappeared; child psychologists refused to read this character and the formative scenes in which she was present. Labor historians and scholars of maternity have tended to follow suit.[25] Filling the gaps, the pauses, and the silences of the literatures on domestic labor, on family life—all of which implicitly reify the sentimental importance of the mother (white, largely middle- and upper-class) over and against other mother figures or caregivers—we can bring the nanny into further view as embodying both the problematic of impression and contagion ascribed to mothers *and* the demand to stand in ascribed to media in the long twentieth century.

The etymology of nanny is as fascinating as it is unsettled: perhaps coming from terms for other proxy-mothers (aunt or granny), perhaps from early sounds infants make in babble. As we might use it now, the term "nanny" came into common usage in the aftermath of World War I in Anglo-America. Previously, this kind of carework was termed "nursing"; and what we now might call a nanny was a wet nurse, a nurse, or, for older children, a governess. On the one hand, each of these designations might delimit the training and subsequent role of the employee in the home, especially when referral services took the model of the British domestic worker (where nannies were trained and graded on a scale well into the twentieth century).[26] Referral services might conjugate the relationship between employer and employee (now this labor is sometimes ceded to platforms) based on a match between criteria and experience. On the other hand, while there were attempts to professionalize the labor of the nanny via accreditation and standardized training in the early twentieth century, with the exception of hospital programs, like that of the Babies Hospital in Newark, New Jersey, very few such programs established themselves.[27]

In her different guises, the nanny also invariably lived in. The proxy-mother was always close to hand in the same home (if in very different parts of it) as the mother. In the twentieth century, she was distinct from ostensibly new and casualized counterparts: the mother's helper or the babysitter. I use "nanny" throughout this chapter to designate the regularized, non-kin, paid, twentieth-century worker called in to care for another's children. "She" helps articulate shifts in carework. Through the era of her assumed employment, and as she gives way to nonhuman intermediaries, we can see how fantasies of progress belied regressive, nostalgic designs. As Lynn Spigel writes, "while technology advances [white] domestic ideals stay the same."[28] That ideal here is one in which proxies aid in maternal labor—whether human or nonhuman—while that necessary aid is selectively occluded.

If the nanny poses a particular problem for the bourgeois family, it is a problem that is suppressed, negotiated, yet constantly resurfaces. Whereas in the eighteenth century and prior, the fungibility of mother (as source of nurturance and care) was implicit in many child-rearing practices, in the nineteenth century, new anxieties about the nanny as site of contamination of the nuclear family and interference with that family's culture came to prominence. While employing a nanny was far from a ubiquitous practice at any point in the twentieth century—the decision to *hire* a para-mother is deeply classed and raced—the prescriptive force of the upper classes moved beyond the boundary of their own homes. Although more often a figure rather than a pervasive reality in middle-class homes, upper-class panics about children, and the nanny employed to care for them, bleed beyond that classed zone of exception and even beyond the nanny—to the daycare and the school (as in the Satanic panic; more about this below). These panics were especially distilled and amplified beyond their own immediate contexts, on the one hand, among the aspirational and, on the other, in those subject to the policing of norms.

The trouble with the figure of the nanny has remained largely static. She is understood as coming from without but moves within, permeating the home with difference even if she lives full-time in the domicile that employs her. She also contains too great a similarity to mother. As nannies and nursemaids themselves testified in the interwar period, this alien quality ascribed to the nanny—that she comes from beyond the family and its "kind" to help maintain it—is precisely where the perceived trouble begins for her employers.[29] While in upper-class homes, childcare is always distributed and might be carried across mother, nanny, and mediated occupation and care (in the form of children's work, furniture, drugs, and disciplinary violence), here I turn to the particular anxieties attached to the nanny, a person paid to cross class lines in order to provide a form of labor in which waged task and maternal care necessarily blur.

The dependency on the extra-familial caregiver has been linked to extreme forms of violence against waged and enslaved care workers who did the work of mothering for children not their own, sometimes while caring for their own children. On the nineteenth-century plantation, as historian Stephanie E. Jones-Rogers shows, the enslaved wet nurse was purchased by white women to secure "the appropriation of their breast milk and the nutritive and maternal care they provided to white children. The demand among slave-owning women for enslaved wet nurses transformed the ability to suckle into a skilled form of labor, and created a largely invisible niche sector of the slave market that catered exclusively to white women."[30] Social, economic, racial, and ethnic difference between mother and nurse (and later nanny) were what made the outsourced care possible but also what elaborated it into a prime site for contamination panics.[31] On the nineteenth-century plantation and beyond it, breast milk specifically became suspect: Might being nursed imbue the child with the qualities of the body it turned to? This panic existed far beyond the antebellum South in this period and its precursors. As Jones-Rogers argues, "bodily fluids and a child's ability to imbibe moral and racial essences through a woman's breast milk . . . served as the basis for stern warnings to new mothers about putting their babies at the breasts of strange women."[32] "Mother as medium" can be traced all the way back to this understanding of maternal impressions; like a mother, a wet nurse would impress herself upon an infant as well, letting in what "should be" kept out. On this understanding the child is also supposed to be protected from foreignness at this level, not just protected by the family structure and walls of the home. Human milk becomes configured as a biological "intruder," not just a medium of contamination.[33]

Not only would wet-nursing in infancy "pass" qualities from the nurse to infant, but her whole person, her attitudes, and the contents of her very labor might be transfused as well. Middle-class parents continued to employ nurses and nannies for their children during Reconstruction, but what a nanny should be and who she should be became narrowly idealized just as fewer "acceptable" candidates (here coded as American-born white, middle-class, unwed women) for service positions were available.[34] This only intensified into the twentieth century when conceptions of breastfeeding recursively swung on a pendulum from the most important source of maternal care (a pure, private circuit between biological mother and child) to remediation by other new technologies, like formula, that met long-standing technology, like bottles.[35] Depending on political conceptions of motherhood as they intertwined with scientific understandings of nutrition, formula could be seen as guaranteeing purity or, paradoxically, itself a site of contamination. (In our present, parents who share human milk, as in the most recent formula crisis of 2022, are also seen as suspect even as networks for sharing milk have a long history.)

Beyond nursing, notions of contamination from childcare workers continued through the earliest years of the Progressive era. One set of worries centered on what the nurse might pass on to the child. Another orbited intently around what nursing might do to the mother-child relationship. As one parenting manual from this moment put it bluntly: "nurses and governesses intervene to cut off sympathy between mother and child."[36] As Lynn Weiner writes, there were also contentious social debates as the "[p]ublic discovery of the young working woman mirrored collective anxieties about changing gender roles, moral purity, the displacement of rural tradition by urban culture, and the fate of future generations."[37] The anxiety of *future generations* is a tell: the nanny might corrupt and corrode the bourgeois family such that, as it reproduces itself, its makeup mutates and degrades. Mother was the fit medium—so long as she could manage her children—and the nanny was the ill medium.

This purity/impurity dichotomy—or, in other terms, immediacy/mediacy—appears in all genres of parenting information in the Progressive era. First-wave feminists began to make the argument for delimiting white middle-class women's work to mothering, relying on early automation (eliding human help rather than excluding paid care) to take care of *other* feminized social reproduction, or domestic labor in the home. Feminists from across the political spectrum have argued a version of this formulation since Catherine Beecher and Harriet Beecher Stowe's 1869 work on domestic efficiency, *The American Woman's Home*. Freed of cooking and cleaning, the Beecher sisters argue, women are better mothers and wives—and therefore need not seek paid help beyond the home.[38] The home, too, exists in an attention economy; remove the burden of some forms of duty, and greater satisfaction and human-to-human connection is possible so long as we allow our devices to automate *other* labor practices. Put another way, automating some tasks would allow a mother to provide not just presence, but undiluted attention to her child. The figure of the nurse interrupted that pure maternal attention and combined with panics about contagion. Seeking out new forms of aid that extricated the family from these problems, parents at the turn of the twentieth century turned to new technologies.

Scientific mothering consolidated this, furthering medical surveillance of mothers and the rigidity of feeding, bathing, and sleeping schedules. This rigidity, as we will see, comes on the heels of the second industrial revolution, and begins to conform the nonconforming rhythms of infancy to the timetables of waged work. At the same time, as these traditional mother's helpers—human, disciplinary, and pharmacological—became suspect and guilt-inducing, innovation and automation offered themselves as emancipatory alternatives. Furniture and child architectures became sites of passive control and care.[39] As the historian of material culture Karin Calvert recognized, in this moment, "parents seemed to want more from a piece of furniture than simple containment. They were looking for something that would exercise and entertain their babies. . . . Essentially, they sought a device that could replicate the attentions of the nurse."[40] If mother could be nearby (providing presence) and the furniture itself could provide attention, together they could cut out the additional, "intrusive" caregiver who was, in most cases, the primary caregiver. In this scene, mother would be released to do the rest of the work allotted to her. This revolution in childcare furnishings—taking the long-standing cradle and adding new affordances to it—accounts for most of the objects still routinely in use today: the playpen (containment and entertainment), the high chair (containment and nourishment), and the cradle's replacement by crib (containment and rest; see the following chapter).

For white middle- and upper-class mothers at the turn of the last century, bringing another human influence into contact with their child was full of xenophobic, raced, and classed anxiety about extra-maternal impression—from the kinds of stories a nanny might tell to the kinds of facial expressions she might make.[41] The mother was then offered—by industry, by women's magazines, by pediatric advice—an alternative. She could be mother, nursemaid, and sentinel at once to circumvent this supposedly dangerous influence—to keep the mother-medium free of the nanny-medium.[42]

Other humans were not the only solution to caring for children that had recently come under suspicion: corporal punishment, opiates, and alcohol had all fallen out of favor by the turn of the century.[43] They too were understood to be instructive or disciplining or pacifying—to help with the labor of raising children—but each met basic criticism for their side effects. If the nanny could transmit her home environment (first in milk and later merely via presence), and other tools were understood to be improper, then the white middle- or upper-class mother would have to constitute the entire alternative, the necessary assemblage. She was to be an environment unto herself.

INNOCENCE IN THE NURSERY

To account for these changes, we need to look not only at the shifts in raced, classed, and gendered conceptions of "proper" childcare, but to children themselves, whose status underwent massive revision in this moment. Why might it matter to a nation that a rich child was kidnapped from his nursery? Only decades prior, it might not have. But by the 1930s, white children were, as sociologist Viviana Zelizer puts it, "priceless."[44] When the category of the mother changes, and when the category of maternal labor shifts, so does her putative object of care. The two co-produce these meanings ascribed to them as individuals; they are entailed upon each other.

The category of the child has of course long been political and exclusive, if not for as long as we might think. To be granted the status of child in the nineteenth century was to be granted humanity.[45] As historians of Black girlhood have shown us, childhood and children were rife with contradictory racial meanings. Robin Bernstein explains that in the late nineteenth-century United States, "white children became constructed as tender angels while black children were libeled as unfeeling, non-innocent, non-children. . . . At stake in this split was fitness for citizenship and inclusion in the category of the child, and, ultimately, the human."[46]

As scholars such as Zakiyyah Jackson, Kelsey Henry, and Mary Pat Brady show, the category of unfitness for citizenship and personhood was also described via a permanent childhoodlike state, akin to disability or animality, and sometimes all three states operating contiguously.[47] In the meantime, older insistences on the productive child transformed into a new characterization of the priceless one. As Zelizer chronicles, this consequential shift changed mainstream parenting expectations and protocols of child-rearing (a shift from children in the street to more time indoors and, when beyond them, a tighter regulations of public space) and occurred alongside a precipitous drop in the birth rate.[48]

This new, priceless, white American child demanded particular safeguarding, just as the category of pricelessness remained exclusive. Implications for the working class are debated by sociologists. Beyond the middle- and upper-class white families where baby was now understood to be a precious individual, as M. E. O'Brien writes, "capitalism had already destroyed the working-class family" across the color line.[49] Zelizer argues that, by the 1930s, this version and vision of innocent and priceless childhood had been extended to the working class across difference, though the Great Depression would restore the need for the working child in working and some middle-class families.[50] In the subsequent hundred years in parenting and child-rearing, the many meanings of family, childhood, and mothering remain deeply inflected by this stark division, wherein a version of idealized, protected, white childhood initiates panics that, as we shall see, license the further surveillance, capture, control, and eventual destruction of other families.

By the time of the Lindbergh baby kidnapping, then, the nanny had come to pose two kinds of threat to the fantasy of the emergent nuclear family complex. The first threat is that, configured as labor and as a necessity, the nanny puts pain to the lie that the family is all that is needed to reproduce itself. The second is that, as an extra-maternal figure marked via difference, she threatened to breach the domestic, bringing with her all those forces (classes, races, origins) that were more properly external. In this tremulous and toxic setting, the relationship between nanny (or nurse) and mother was configured as one of uncanny doubling and explicit difference. Exactly because the figure of the nanny gathered to itself all the work of dispersed mothering, and exactly because she was a single figure, a mother was now defined by what she is not able to offer, not able to convey. When she is not present and offering attention, someone else—or something else—is assumed to be. In the longer history of paid (and unwaged, enslaved) care proxies, the nanny was thus both a replacement for the mother and a medium in her own right—much like a mother herself. She straddles the conceptions of both object and subject in the long twentieth century: she is both a tool for care (a proxy, a form of technology) and a maternal medium rife with her own possible contaminative influence.

THE NANNY ON THE COUCH

By the time of the Lindbergh baby kidnapping, one nursemaid could write with some worry that she was one of a "disappearing type," and nurses' unions began to form around work protections in response.[51] The psychological sciences were not immune to these historical changes, even as the employment of nannies began to disappear from the American landscape.

In the literatures produced by child psychologists, the archives of the psy-ences with which this book is especially concerned, the nanny was everywhere but also nowhere. While sometimes being accorded the status of the mother's double, from the Oedipus complex forward, she is not treated as if she has the same formative impact. At the same time, she is absolutely understood as impressing herself—often literally, sexually—on her charges. In other writings, the governess, the nurse, and the nanny are mentioned, assumed, gestured at in a half-clause and sometimes outright forbidden in direct recommendations to parents (made present by her absence). In the technical papers of the major psychoanalysts and psychologists of the twentieth century and in their address to their parent-readers, the figure of proxy mother thus occupied a strained and peripheral position.

A consideration of implied and explicit theories of the nanny in the psy-ences requires us to return to the end of the Victorian era, and to Sigmund Freud (who was being widely read and disseminated by himself and his followers in the interwar period). Freud had had a beloved nurse of his own. She was a Catholic Bavarian, and Freud's ambivalence about his own Jewishness is often said to originate from his deep love of her. (Freud apparently once asked his mother "whether she remembered my nurse. Of course, she said. She was always taking you to church. When you came home, you used to preach and tell us all about how God conducted his affairs.") Also salient: the nurse allegedly stole from him and was sent to prison when he was a toddler. The beloved mother double disappeared.

Once Freud grew up and began telling us how we were conducting our affairs, and conducting his own, his interest in the figure of the nursemaid and governess returned. Beyond his personal biography, the governess as a figure haunts Freud's work. The governess is at the center of Freud's theories of hysteria (Lucy R., one of the women in *Studies in Hysteria*, is, indeed, a governess; Freud accused his famous patient Dora of acting *like* a governess when she terminated their contact). Contra the notion that Freud only worked with the wealthy, particularly before World War I, Lucy R., was an exception to this false rule; the figure of the governess and her colleague, the nurse, are invoked in dreams, in jokes. And of course, both nurse and governess were caring for Freud's children, and had cared for both analyst and many of his analysands in childhood and youth.

Freud was opaque to himself about his thoughts on paid caregiving—even as the figure of the psychoanalyst is one of waged care. He thought of the governess as a mother double, one who played actually a more central role of psychological mothering, especially by the nurse in infancy and early childhood. Freud was deeply worried about the seductive power of the nanny. As psychoanalyst Daria Colombo writes, this was particularly centered on boys, although girls were not considered immune to it. Crucially, whereas seduction by parents was understood as fantasy (after 1906), "it is the boys whom Freud consistently described as having experienced actual seductions by maids, seductions clearly delineated off from fantasy by Freud."[52] Rather than an obscure, biographically specific panic that Freud then universalized, as historian Carolyn Steedman argues, "The domestic service relationship was one of the most widely experienced in the society" in which Freud was living.[53] But that Freud saw the figure of the nurse as seductive, or "an unending source of sexual excitation," as he wrote in 1905, was indicative of how the mother's double was figured psychologically.

Yet Freud both elevated the figure of the nurse—as the person who teaches children sexuality—and missed her relevance to some patients completely. When a governess was of central concern to a patient—her comings and goings, her departing the family's employ for marriage—and she brought her memories into treatment, Freud remarked, "I could make nothing at all of these reminiscences."[54] Where he was able to make something of the governess, it is purely as category, as figure, as a universal. In his early work, when he thought neurosis was related to widespread abuse of children, Freud turns the governess or tutor into a figure of sexual initiation and trauma.[55]

Even so, psychoanalysis, with its great *pretense* to universality, made a telling slip. Once it presented a more unified theory of child development, under the sign of the bourgeois Victorian family of fin de siècle Vienna, families were largely represented as living in a way they did not: alone. Through a psychical, developmental sleight of hand, Freud rewrote psychic history as though almost all medically *relevant* domestic and child-rearing labor occurred within the family, a self-reproducing unit. The nurse and the governess might be seductive, might be the very *origin* of neurosis in Freud himself, but actors equal to parents in the Oedipal drama they were not. Freud's instance on the governess as the matrix where children learned sex and were seduced is in keeping with the panics being experienced some 4,000 miles away at the same time in the United States—and again and again across the twentieth century.

This changed with the advent of behavioralist *and* child psychoanalytic approaches to the child in the late 1910s and 1920s, and when the middle- and upper-class family—and its employ of servants—suddenly transformed. As we have seen, a drastic decrease and removal of waged domestic workers had already taken place in the interwar period, including waged childminders.[56] The Great War had shuttered borders on both sides of the Atlantic, and although they theoretically reopened thereafter, there were mass immigration laws, which thus, especially in the US context, as Ruth Schwartz Cowan writes, "drastically reduced the influx of foreign-born young women, who had previously constituted the largest portion of the servant population. In addition, the expanding economy of the 1920s increased the opportunities in factories for women without skills. . . . [T]he business-class wives of Muncie reported that they employed approximately half as many servant-hours as their mothers had done, at roughly five times the wages (salaries for people in the middle-income ranges had roughly doubled in the same period)."[57]

This had immediate consequences for psychoanalytic and psychological thinking about the nanny. In 1919, behavioralist John Watson, together with his graduate student Rosalie Alberta Rayer, turned to conditioning infants at the Harriet Lane House at Johns Hopkins, which Leo Kanner would go on to convert into the first clinic in child psychiatry (see chapter 3).

In images of one experiment, a little baby swaddled in white sits on a thin mattress between John Watson and Rosalie Alberta Rayner. The baby, given the pseudonym "Little Albert," was just nine months old, and although controversy remains over his identity, one of the stronger hypotheses holds that he was the son of a wet nurse employed at Johns Hopkins Hospital—showing the absolute vulnerability of mother-doubles who were mothers themselves, well into the twentieth century. Little Albert was selected to be conditioned to fear: rats, fluffy white things, and even Santa Claus. This 1919–1920 study had two purposes: one, to supply evidence for Watson's idea that three emotions—love, fear, and anger—are inborn and can be stimulated; this would become a crucial part of his work in *Psychological Care of Infant and Child* (1928), the US parenting bible until 1946 (when Dr. Spock took his place). The second purpose was to show that those innate emotions and their stimulus could be transferred or, in Watson's parlance, *conditioned*—which mothers could then replicate at home, sending the clinic back to the parlor.

Little Albert was not afraid of many things on the first day Watson and Rayner observed him, not of rats or rabbits, but he was afraid of loud noises. Watson and Rayner then tried to transfer his fear of noise to those objects he was not previously afraid of. Watson and Rayner met with Little Albert for months, sequentially trying to condition his fear response. Watson claimed they had successfully conditioned these fears, even as the results could not be replicated by their peers. The study had another consequence: during it, Watson began an affair with Rayner, which was discovered by his then wife. We can see that Little Albert was given his case name so as to be the masculine diminutive of Rosalie *Alberta*—the subject become their fantasy baby who, upon being taken from his biological mother, might be reared according to their scientific dicta. Once the affair was discovered, Watson was ejected from Johns Hopkins (rumors followed—and have since been largely discredited—that he was fired not just for the divorce, but because Rayner and Watson were turning to the field of sexology, studying their own sexual responses to one another).[58] When it came time for Watson to address mothers publicly, Watson was a champion of mother-only care. As he wrote in *The Psychological Care of Infant and Child* (1928), a perfect home was one in which there was no help—even if the "help" is from whence baby originated.

After the Great Depression began, there was a further drop in the employment of servants in the upper classes—including governesses. As Cowan details, by 1931, when President Hoover hosted a Conference on Homebuilding and Homeownership, only 17 percent of respondents reported that they employed a servant—not likely to be a nanny.[59] The nanny disappeared from psychoanalysis in this period as well. In 1939, the pediatrician and psychoanalyst Donald W. Winnicott, who made a career both in private practice and in addressing mothers directly by radio broadcast, declared in a meeting a key formulation for thinking through the relationship between mother and child: "There is no such thing as a baby"; there only exists a "*mother-baby.*"[60]

Winnicott here means to affirm the mother-child dyad. He gave us an articulation of it as always interlinked, relational, co-constitutive. We might say that Winnicott conceived the mother and baby as mediating one another. In this way, Winnicott was radical for his moment. He introduced not just the baby's dependency on its mother as a medium of care, but the notion that the baby is in a form of mutuality with the mother too (this would become a hallmark of feminist psychoanalysis some thirty years later). In affirming the first supposition—that the mother is a total environment—his thinking remained typical of the mind sciences in this moment; in the recently consolidated field of child psychoanalysis that had stunningly taken shape in the preceding decades and in nascent pediatric psychiatry, the mother is understood to have near-total influence on infant and child outcomes (which reversed Freud's emphasis on the father; see introduction). Like all works centered on pathology, the outcomes that were addressed, it must be said, were the "bad" ones; Winnicott's work is no exception. His suite of essays on mothering expresses the depth of the demand on mothers, one he refines and makes his own. Winnicott, describing what he calls the "good enough mother," seems to lower the bar from perfection and is thus often misread as proto-feminist. Recalling Spock's notion of a mother being able to trust herself, but with a twist, for Winnicott, mothers must be deeply attuned, cannot be trained or taught to be so, and cannot merely go through the gestures of care (he cautions against boredom during diaper changes).

This good-enough mother was always to be coupled with her baby. Winnicott, misinterpreting his contemporary for the detemporalized universal, was largely correct in offering this description of mother-baby, though not exactly for the reasons he articulated. That Winnicott ventured mother-baby and not carer-baby was no accident. Fathers rarely appear in the work of Winnicott (or other maternalist work of this period),⁶¹ and he is additionally addressing a UK context wherein, much like in the United States, there had already been a drop in domestic workers—including the figure of the stalwart British nanny.⁶² Although Winnicott treats this as a generic phenomenon, one existing out of time and society, he is nonetheless reacting to massive social change. No longer were there to be proxy mothers; for baby to exist, baby would have to be in pure relation, so argues Winnicott, with its mother.

Though describing the British mother-baby, this formulation was championed by clinicians and some feminists on both sides of the Atlantic and provided a new name for a long-standing conception of the bond between mother and child. The phrase was so apt it seemed at once both intuitive and commonsensical.

At the same time, as he took his idea further, Winnicott's innovation gradually began to stagger under its own weight. He writes of the baby, "[T]he unit [of the baby] is not the individual. The unit is an *environment-individual set-up*. . . . It is in the total set-up."⁶³ Winnicott continues: "if you show me a baby, you certainly show me also someone caring for the baby, or at least a pram with someone's eyes and ears glued to it."⁶⁴ Winnicott's "total set-up" focused on the mother-baby, and how it, as a unit, gets on. Winnicott increasingly blurred the edges of the mother, splicing her with attendant technologies, enlarging her. But he affirmed the exclusive interdependence of mother-baby while also immediately showing us that it requires an array of forms of care—now a fungible term referring also to labor. It is as if Winnicott is saying, as a contemporary scholar of technology might, "we are nothing independent of our socio-technical assemblages." If there's no such thing as a baby because a baby needs a mother, there is also no individual, simple thing as a mother. She is never just herself. And she is not only conjoined to baby. To be mother-baby, the *constellation, the matrix*, the "total set-up," or the medium, must include, at the very least, a pram and someone attending to the baby inside it.

In imagining this intersubjective mother-baby, Winnicott is actually describing carer-baby-technology. Winnicott does not recognize any contradiction. Rather, he naturalizes these low-level, ubiquitous technologies as part of a setup (and hides away the nascent welfare state in the United Kingdom, and the breadwinning father assumed to provide for mother-baby). Writing in an era in which a mother is supposed to remain an "environment"—a term that would be highly associated with Winnicott although not originated by him (see the introduction to this book), because she is assumed largely to have produced the infant's reality through her body, as well as its original environment in pregnancy—Winnicott made an interesting extension. He tacitly included an entire second infrastructure, a material environment. And he called them by the same name: the psychical-maternal and physical tools all become part of the maternal environment, a heterogenous and porous matrix—a medium—paradoxically considered exclusive and inviolate.

How exactly did the lone, ostensibly middle-class, white mother with her pram get thought of as a baby's medium? Doing so entailed training eyes away from the irreducible diversity of families: single-parent households, multigenerational households. It involved ignoring immigrant patterns of family-making and Black families' survival strategies to elevate and naturalize the mother-baby living on a family wage instead, in the UK context. It meant suggesting that there is something innate, instinctual about mothering, and that she is always the right medium—an expansive, extended medium—of care, but never the right version herself (at best she's "good enough"). She is configured as encompassing activities and objects, including her labors and her body, her temperature, her moods, her attention. In her expansion to a field and reduction to a function, she retains her singularity, not so generic as to be interchangeable with another person. This a double bind: at one level the discourse says that mother is fungible—replicable by technologies or media; but any mother who takes seriously this fungibility and replaces or prosthetizes herself with another is degraded.

Winnicott was writing at a time where, especially in the UK context, most of the leading analysts were women (although there were plenty of men in positions of power and often in charge of analytic and psychological societies). This was the same moment when the analyst herself became reconceptualized as a para-mother via transference ("remothering" the patient). Nonetheless, the nanny remained theoretically scarce. When women became the major theorists of psychoanalysis, the figure of the nurse and governess (or later, and in England, the nanny) became completely repressed.

The very children seen in the analytic consulting room would have likely oscillated between two sites of primary care (as Anne McClintock argues, this class of child experienced growing up with a nanny-mother structure as having two mothers).[65] Again the mother and the nanny often straddled the internal, familial, and the external, a divide marked by difference if not racialization. Melanie Klein hid the possibility of a nanny in her footnotes along with her bottle feeding, placing the figures who replace the mother and the breast beyond the ken of normative development.[66] As Emily Green shows, this split between mother and other, literalized in Klein's texts, was highly racialized for the theorist, which Amber Jamilla Musser has extended to show that Black children, as well as Black mothers and caregivers, have been ignored by psychoanalysis.[67] D. W. Winnicott excluded the nanny entirely and only mentions the nursemaid once directly—to make her function equivalent to the analysts in his paper on the "false self."[68] Jean Piaget, the Swiss psychologist, only paused to think about his nanny because when he was fifteen years old, she wrote him a letter to say that she had lied to him: she had said she once protected him from a kidnapper and there was no such event.[69] Because Piaget had made a memory of this scene, the nanny's only relevance was to his studies on memory.

Waged caregivers then are figures whose central psychoanalytic repression might be read as a symptom of the theory itself. The theory's willful failure to register the nanny, to account more fully and complexly for her place in the libidinal economics of the home, can, following Gillian Strecker, be read as recapitulating the psychical experiences of dehumanization.[70] After Freud, the psychological impact of the nanny is merely gestured at—that the beloved proxy mother might expand our conception of why children growing up in upper-class families develop the way they do—but *how* this might be the case is almost never investigated until the early twenty-first century. The governess is a presence, she is a fact, but she is always glossed over. The other dominant branch of child psychology in Lindbergh's moment, behavioralism, even more directly commanded its followers: good families must not have nannies.

TRANSMITTING CARE

By the moment of the Lindbergh baby kidnapping, as a 1930 *Good Housekeeping* article dramatized, the dying generations of nannies were misguided in thinking themselves essential, beloved. The nanny might be devoted, but she was a fool: children quickly forgot her as soon as she left, undermining the tradition of nursemaid or nanny caring not only for the children of a family but for a subsequent generation as well.[71] Told through the parable of social change, "pert flapper nurses" (themselves more in vogue in the previous decade) and rude teenage girls who no longer wished to wed (the traditional ending for a nanny's employment), the nanny was understood to belong to the past. If not for social reasons, then technologization of the home was understood as the reason for her dismissal. Yet nursemaids themselves knew that the introduction of time-saving devices was not likely to obviate or automate their work—even if other domestic workers had cause to worry about new demands and skills placed on them. Instead, the threat to the nursemaid's employment came wholesale—replaced by crib or obviated in the middle classes by the rise of the nursery school in the twentieth century.[72] By retiring her as relic, even the panics surrounding her role in the home were repressed.

This change was documented by the nurse-maid-sociologist Martha Haygood Hall in her dissertation, "The Nursemaid: A Social-Psychological Study of the Group," submitted to the University of Chicago in 1931—just two years before Charles Lindbergh Jr. went missing. Hall, who had worked in the city before becoming a student, gathered her data via, in her words, "acquaintance with" other caregivers, both in the course of her employment and through interviews with the local YMCA group for nursemaids. Her contention was that to be employed as a nanny presents the worker with a double bind: the wages offered delimit intimacy by making it work, yet that work is literally to dwell and act within the intimacy of a nuclear family. Turning her attention to "nurse girls," "nannies," "mother's helpers," "governesses," and other forms of paid childcare, Hall determined quickly that it was the individual's life station—their race, class, education, training, and possibly connection to the family—that determined their title as much as years on the job.⁷³ But the nursemaid—the nanny—almost without exception was employed by the upper class and represented the dominant mode of childcare *for* the upper class.⁷⁴ Uniformed nursemaids were the norm; upper-class mothers who *mothered* were noticeable.⁷⁵ They were the exception that proved the rule.

Given the classed and raced panics around nanny-as-medium, were she to be employed by the white upper classes nonetheless, she must be monitored. As one nursemaid in the period put it, "You can't be a servant and be a human."⁷⁶ This too inflected the meanings of motherhood for those who worked with and employed nursemaids: she was termed "parasitic," "idle," and had "dried the springs of mothering" by refusing her duty.⁷⁷

Given that the nanny was under intensifying scrutiny in this period, and that she was already herself understood as a problematic medium of care, it makes sense that inventors and tinkerers might try to turn nanny, or at least some of her activities, *into* media. Just five years after Charles Lindbergh Jr. went missing, the president of Zenith Radio Corporation, Commander Eugene F. McDonald Jr., commissioned the first baby monitor, improbably designed by a famous American sculptor, Isamu Noguchi.⁷⁸ McDonald was also quite rich, and a father to a young child. Worried that his daughter was a prime candidate to be the next Lindbergh baby, he needed a device that would afford him a security the Lindberghs lacked. A human staff was clearly not enough to safeguard his little one. A perfect mechanical listener became a must, a replacement. The Zenith Radio Nurse was first to be used aboard McDonald's yacht, where he, his wife Inez, and their daughter made their home; he explicitly wanted the couple to be able to put her to bed securely at one end of the boat and entertain at the other without sacrificing knowledge of her, allowing Inez to host and mother simultaneously—much like the turn-of-the-century mother was supposedly able to work while her baby was in the arms of the mechanical crib (see chapter 2).⁷⁹ Under commission, Noguchi rendered the device's two parts, the Radio Nurse Receiver and the Guardian Ear Transmitter, in Bakelite and metal, respectively. The distribution across two devices—one standing in for care and the other for surveillance—mirror the dual charge for mother to be attentive *and* present in this moment. As much sculptural innovation as domestic technology, the

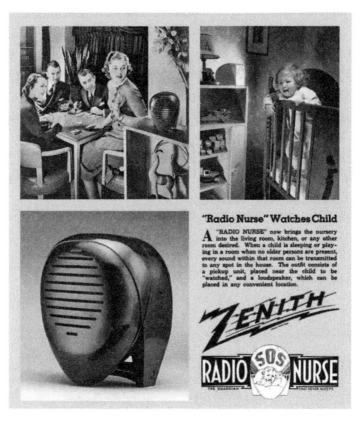

Figure 1.1
Advertisement for Zenith's "Radio Nurse," Artist Rights Society, New York.

Radio Nurse appeared in the 1939 Whitney Museum sculptural exhibition and was found to be "prettier as a radio than as a nurse," as a reviewer of the show writing for *Time* magazine put it.[80]

Noguchi's collaboration with Zenith, though mass produced, was a failure. The radio had serious technical shortfalls, including its operating on the same frequency as other domestic technologies, which made it prone to static. Nonetheless, the idea for the device lived on and was improved. Other related forms of baby monitoring were soon on the market, all building off the basic principle still in use today: place a transmitting monitor near a baby and a receiver near the parent. Early users bought into the fantasy and promise of extended parental attention, of having total knowledge of the nursery, in addition to their fascination with the actual functioning of the device. If we look back at the security concerns McDonald had (yacht notwithstanding), we can see that our most extreme fears (kidnapping, death) cover over entrenched fears around the nanny, and the increasingly common babysitter, as mobile, non-nuclear figures.[81] Zenith's designers chose telltale names for the receiver and transmitter: the Nurse and the Guardian. Not only was the Nurse not alone, but these two roles encapsulate this notion of augmented parental function that allowed parents to strive toward an ideal of control via technology, delegating responsibility and then surveilling the delegate. The Nurse and Guardian tend the baby when the parent cannot or does not need to actively—during sleep. As a mechanical device, it was not at risk of falling asleep itself, nor could it intrude upon or harm the baby while working under the sign of care. It is inhuman, perfectly vigilant, and without influence.

The nanny as figure was not the only worker—waged or otherwise—whose paradoxical threat *and* necessity required remediation. To synthesize the paradox via automation was to retain the fungibility of workers while solving for the problem of their power to undermine the family. This often presented as the turn to mechanical people as much as it did devices bearing their name (and it is no wonder that we have thus lived with a century of panics that robots might rise up—just as workers have done and do). Catherine A. Stewart details a particular version of this fantasy among "white Americans for a permanent labor force consisting of unpaid and unresisting black underclass . . . [which] reach[ed] its apogee during the Great Depression" in the "Mechanical Slave," produced in the same moment by Westinghouse Research Laboratories.[82] The "humanoid robot," a time-saving device with very limited capability, was a horrifying vessel of white nostalgia for slavery. What mattered more to its creators was its symbolic power, not its use value.[83]

Whether a machine that directly remediates the enslaved worker or Noguchi's remote parenting device, the goal was to automate out a possible contaminating influence while installing necessary labor. Noguchi's gendered couple, the Nurse and the Guardian, describes the two main functions of parenting—nurture and protection—and expresses the techno-optimistic fantasy of control emerging in the 1930s that parents still contend with today. They also reveal that a nurse alone is not to be trusted; the guardian—a legal term for the duty to provide and care—must be omnipresent.

MAKE WAY FOR SURVEILLANCE

The postwar economic miracle only further entrenched ideals of the nuclear family and its self-containment. As Emilie Stoltzfus writes, "This was a moment of great flux for women's paid labor, and yet, 'the stay-at-home' mom remained in the majority. Both popular writers and historians of women have noted the substantial barriers in this period to the employment of women, especially mothers of young children. These included the dictates of popular culture, professional psychiatric, psychological, and social work wisdom, and cold-war political goals."[84] In the UK, even attachment theory was used as a cudgel to close war-time day nurseries (see chapter 4).

The babysitter was the key emergent figure in the prosperity that followed World War II. As historian Miriam Forman-Brunell shows, "the sitter" was a highly visible source of middle-class anxiety.[85] While the sitter did not *replace* or replicate mother, and did not interfere with the notion of staying home in the day to care for house and child, the figure recalled earlier, prewar nanny panics in a new form. The babysitter was both sexualized and seen as immature—doubly suspicious. But this was an ad hoc form of care (and an ad hoc form of employment). It was to be used sparingly, to reaffirm the marriage. Rather than replacing nanny, to contend with other domestic duties, mothers increasingly left baby and child alone—*contained them* and then surveilled the area in which they were contained.

Mothers securitized the home by adding special furniture with reporting devices. Even though these devices were not wireless, and could not transmit their surveillance to the mother elsewhere, they were meant to provide both a feeling and a reality of security. This was seen as a new, pervasive, and erroneous way to manage the family in the middle class. One 1950 *Good Housekeeping* article, aptly titled "Never Leave Your Baby Alone,"[86] offered a litany of accidents that might befall an unattended child, addressing the uptick in parents doing just that—leaving the child (especially little children) to be monitored by nothing more than a device. Parents would run errands, ferry one another in cars, or even leave for the evening, trusting the child to their home and its infrastructures alone. The suggestion of *Good Housekeeping* was, of course, to have other humans aid in the work of parenting. In the inventory of possible childminders the advice columnist presented, *none* are paid professionals (the grandmother is suggested, but only if she herself volunteers; other mothering in these white middle-class mid-century extended kinship networks is marked off as largely inappropriate). Mother was supposed to do the work of containing the child, to be its only medium of care—and to do so alone.

Correspondingly, individual theorists of mother-baby interaction continued to use the labor market, and the disappearance of the servant class, as alibi for their early inattention to the role of the other mother. Within the psychiatric and pediatric literature of the time, the nanny began to feature less and less or, when she did appear, was condemned outright. In 1945 Winnicott advised, "The process [of breastfeeding] is immensely simplified if the infant is cared for by one person and one technique. It seems as if an infant is really designed to be cared for from birth by his own mother, or failing that by an adopted mother, and not by several nurses."[87] Mothering was, as historian Ellen Herman shows, undergoing its own hierarchization, with biological tie being strongest and forms of adoption following on. Mothering was understood to be interrupted and confused by the nurse. Proxies, in their plurality and their fungibility, were understood as not only having a negative impact on development but *duplicating work*. The next year in the United States, the radical (and eventually beloved) Dr. Spock essentially foreclosed the use of the nanny in the middle-class home. In his best-selling *The Common Sense Book of Baby and Child Care* he mentions the nanny only once, to banish her as a relic of a bygone era.[88] The mother is supposed to "trust herself"—and no other.

The suburban home was therefore often designed so that children could be both contained in their rooms and observed by mother as she did her other reproductive labor (for more on this, see chapters 2 and 6). For babies, whose place was understood nearly unanimously to be in a crib (even as parenting practices remained far from homogeneous; see the next chapter) and in their own solitary room, baby monitors continued to be a useful device, relinking that isolated space with the rest of the domicile. Noguchi's baby monitor was thus reconceived, and that relaying device moved from radio transmission to image. In 1953, using the very first closed-circuit television device, the Radio Corporation of America debuted its "TV Eye." The baby, under this conception, remained an object upon which to train a mechanical, perfect other. Here, mechanicity reprised coding the machine *as* human, drawing on and reversing the notion of the servant as machine—taking the human eye as metonymy and replacing it with a camera. The private TV Eye rhymed further with the private "I" (investigator), popular at this moment. Instead of having a nanny on staff, the family could employ mechanical security.

The TV, which as media historian Lynn Spigel shows was rife with contradictory meanings for family in this moment (see chapter 5), becomes a space of security, observation, and protection.[89] In RCA's ad copy, feminized figures of care ("nurse") are replaced by the *watchman*, a military term whose resonance might be especially loud for the postwar era. The father is at the controls while the mother is seated. This advertisement departs drastically from the imaged uses of the soothing machines of the turn of the century. Here, surveillance is masculinized and the housewife is depicted as completely passive, merely viewing the screen (she is neither doing housework nor working the controls). Despite this being an advertisement *for* a product, the baby recoils, horrified (and as any parent might know from looking at the ad, a baby would knock the tripod over in minutes to deal with any perceived threat or simply to self-amuse). In keeping with the iconicity of the mid-century housewife, mother is reduced and neutralized rather than emancipated (as we saw in "The Veldt"). As Ruth Schwartz Cowan first called to our attention, machines were actively used to replace domestic servants in the middle-class house in this period, and, "indeed, many people purchased appliances precisely so that they could dispense with servants. It is not, consequently, accidental that the proportion of servants to households in the nation dropped just when washing machines, dishwashers, vacuum cleaners"[90] became prominent features. But what remains missing from this account, and what I seek to add, is how *other* devices were used

Figure 1.2
Advertisement for RCA's "Tireless 'TV Eye," *Popular Science*, August 1953.

to keep childcare providers out of the home. Even in speculative and Googie architecture (the latter a form of futural architecture most commonly found in Southern California in the postwar era) and other representations of the future, from "The Veldt" to the early 1960s vision of *The Jetsons*, the remediated nanny figure (the robotic "aunt" of Rosie) was central to provisioning a fully automated life. In reality, following what Sianne Ngai calls the "gimmick" of time-saving devices, rather than freeing the mother up for other reproductive activities (like sex), all eyes serve the Eye.[91]

Here, the anxiety over tending baby escorts the device, and licenses the increased surveillance of the home and beyond, wherein the child is made equivalent to valuable property. Even if the child is priceless, it is still an asset. Fears of what might come *for* baby and these other valuables are highly raced and classed, tied to segregation and suburbanization, and proto-smart houses in this moment debuted CCTV as a standard feature.[92] Such loving scrutiny quickly extends beyond the crib to the whole home, the street, or the neighborhood. Media regimes in the home—here figured as a new form of child surveillance—domesticate these forms of watching, and morally wash them by making them an extension of the maternal.

THE RETURN OF THE NANNY PANIC

The nanny figure and her automated functions are not only a threat to child; she is, in the long twentieth century, as before, too frequently figured as threat to mother and her purity, an intervening medium that, much like the mother, must be controlled, securitized, and monitored for what she imparts to the child. The threat to the baby is therefore suddenly everywhere, supple, mutating. As we will see across the rest of this book, it can come from inside (a parenting manual, a favorite blanket), from outside (an intruder, a culture), or from someone who crosses the domestic threshold under the sign of care (a nanny or, in reverse, a daycare center). There have been panics about all these forms of danger—some addressed via medicine and pediatrics, some via faulty gadgeteering rushed to corner a market, some via a concrete and nostalgic remediation of enslavement. As I argue throughout this book, whatever the ostensible object of anxiety, these are inherently panics over mediating the child—both media and mother are sites of scrutiny. Parental fear is a near universal, but what we fear is not; the primacy of each threat varies by class and race, personal experience and its intergenerational transmission, and history.

The return of the nanny panic in the 1980s reveals many facets of this logic. She was a waged guest, in contrast to the neighbor who, following Wendy Chun's conception, we might understand not to be a threat, even if antagonism can bloom there;[93] neighbors benefit from homophily. Like begets like and protects its sameness. As a cultural figure without much of a social referent, the nanny contained a perceived biological and sociological distance, one that is both repressed and obsessed over. That difference is reprised in her mechanical remediation, whose sole aim is the conversion of an inconvenient human to a nonhuman. It turns out that the fictive nuclear family cannot reproduce itself without external assistance, and that assistance's nature, human or mechanical, is always a problem.

The brutal truth is that children *are* vulnerable, and this vulnerability is multiple: to their own bodies, to external influence, capitalism, state violence, climate crisis. Dangers are, of course, both material and psychical. Gadgeteering and mediated mothering addressed the parent as consumer and their concerns both real and imagined, but the double bind of augmented parenting persisted beyond mid-century. Assistive technology addressed fears, but its solutions generated new panics around attention, presence, and environment.

If at mid-century, parents experimented with leaving baby alone, with new consumer devices (not just TV Eyes but other closed-circuit surveillance) to watch baby, the overt panics around the nanny subsided somewhat in favor of discourses centered on the mother herself, as we shall see in the remaining chapters of this book. But with the surge of married, typically white women working beyond the home—reaching its stabilizing number of three fifths of women of working age in the late 1980s and early 1990s—and in the aftermath of the Women's Liberation movement's insistence on childcare, several repressing, conservatizing childcare panics returned once more. The contradictory notion that caregivers are both proxies and susceptible mediums for child-minding, especially after the Satanic panic of the 1980s (wherein many childcare providers were accused of being members of cults, sometimes cannibalistic ones), renewed worries over selecting daycares and nannies, and what might happen when parents could not supervise such care. Parents also have had to worry over the ordinary flaws intrinsic to child-rearing and caring showing up in their para-parents—exhaustion, forgetfulness, and even anger—as well as what happens when these normative states turn lethal. Nannies, like the caregivers at the center of the Satanic panic in this moment, are liminal—they come from outside but act as parental proxies. They need to be trusted and yet this is understood by the very people who employ them to be a basic impossibility; trust is required absolutely in the other direction, and yet as those employed as caregivers have testified to for the last few hundred years, working conditions are highly variable, and often deeply vulnerabilizing. Not only must employed caregivers contend with contingent work, but even when contracted, when things go wrong, childcare providers are often the first people parents and police look to blame—as was the case in the Lindbergh baby kidnapping investigation.

Patrolling the normal for the lethal was something parents were indoctrinated to do in the 1990s as understandings of child death and parental responsibility grew more complex and well defined, and diagnoses became more particular. From shaken baby syndrome to sudden infant death syndrome (SIDS) to postpartum forms of depression, anxiety, and psychosis, parents are taught about the threats they themselves pose to their children, and how to walk away, regroup, or get help before they inadvertently harm their babies. Mothers are especially psychologically monitored in the aftermath of birth, and before they can take their babies home from the hospital are often shown videos, required by law in some state level, about both best practices and horrific outcomes, or are routinely educated about these potentially life-threatening issues even before their babies are born.[94]

In shifting from analog forms of surveillance to the digital, the "nanny cam"—much like the TV Eye before it—proved remarkably flexible, extending earlier techno-logics of nanny remediation. It takes in its field of vision indiscriminately; only its colloquial name and location (or locations) indicate its primary object of concern. This cute, breezy name hides the fact that it is a camera or even a closed-circuit television placed specifically to monitor an employee doing their job. It also hides a slippage whereby the nanny cam is nannying the nanny—and her charge—at once. It captures images of the nanny to ensure her care; or, we might say, it extends the reach of the employer, again disrupting notions of distance inherent in waged caregiving relations. The drive for extended parental knowledge and the faith that it is a justified means of protection and prevention are nowhere better codified than in the little camera stuffed inside an innocent teddy bear. Instead of monitoring the street or the house next door, the nanny cam turns surveillance inward, to the nursery. As we have seen in this chapter, parents deploy technology to keep their children safe from three general categories of threat: accidents and other forms of unintentional bodily harm, the world outside the home, and the invasion of the home by an agent from that world. The quintessential baby monitor seeks to reassure parents about all three forms at once.

In the 1990s, the media circuit between the nanny cam and the nightly news worked in new ways to reinforce parents' fears over what was happening when they left the house. Judging by sales, it is clear that consumers' will to know, and the hope that knowledge would provide control and peace of mind, led them to the nanny cam—where they reencountered all the mobilizing disquiet they had hoped to leave behind. The evident tensions at work in remediating care, ceding or augmenting parental control, and the moments where media, technics, and societal shifts converge remade the scene of primary care for children. With it, they remade the status and valence of both the nanny and the mother. The story of domestic and parenting technology is that of the supposedly frictionless incorporation of speculative technology toward a vague sense of women's "liberation" inherited from the Women's Liberation movement but without that movement's trenchant demands for better childcare support. The deployment of these caring technologies further obfuscates which carer we are determining as free, and what the afterlives of supposedly emancipatory technology contain.

Shaken baby syndrome is a particularly valuable example of the co-construction of medicine and media, only to be amplified *by* the media. Although first identified in the 1970s as whiplash syndrome and later as whiplash shaken infant syndrome, shaken baby syndrome grew in prominence and codification in parallel with the little machine that documented the violence that supposedly caused it: the nanny cam and the irate nanny.[95] It took the supposedly factual, medical diagnosis and placed the danger outside the family. This was typified in the case of the "killer nanny," Louise Woodward, which galvanized the nation in 1997. In 1997, at nineteen years of age, Woodward was serving as an au pair for the Eappen family in Newton, an affluent suburb of Boston. Both parents worked out of the home. One evening, Woodward brought the baby, Matthew Eappen, into Children's Hospital in Boston where he was diagnosed with shaken baby syndrome—a human-inflicted head injury. He fell into a coma and died five days later. Woodward was subsequently charged with the murder of the eight-month-old boy. The fascination centered, perhaps surprisingly, less on Woodward than on her employer. Deborah Eappen was the emblem of the conflicted mother of the late 1980s and 1990s. Liberated by cultural feminism in the aftermath of the second wave, she worked outside the home. Yet she *only* worked three days a week to keep her career intact; she also reported spending significant time with her child. She "had it all," and then she had nothing. Her life was ripe for being turned into parable. Some members of her community spoke with the press when reporters descended on Newton—telling reporters that they felt Eappen should have just stayed home, solidifying divisions in Eappen's town between working mothers and stay-at-home mothers.[96] The local spectacle that surrounded the case was so intense that Woodward's defense tried to get the location for the trial moved, though to no avail.[97] Beyond Newton, the case revived conventional modern concerns about whether or not mothers should work while raising young children, and if they do, exactly how they should raise them.

Louise Woodward was found guilty of second-degree murder. Later, her case was appealed, and her conviction and corresponding sentence were reduced to involuntary manslaughter.[98] The presiding judge said on the occasion of the reduction, "the circumstances in which the defendant acted were characterized by confusion, inexperience, frustration, immaturity and some anger, but not malice in the legal sense supporting a conviction for second-degree murder."[99] Confusion, inexperience, frustration, immaturity, and some anger: a normal, even unavoidable range of parenting (or carer) affects and states was blamed for the involuntary death of the baby.

Other spectacular and well-reported cases followed, even that same year, including in Thousand Oaks, California, where videotaped images of a nanny abusing a seventeen-month-old were leaked to the press. These images were then used as hard evidence, repeatedly, for criminal conviction. Here we can note a feedback loop between the images of the nanny cam, the press, and the judiciary. Jason Allami, who sold hidden cameras in teddy bears (to be posed as ordinary gifts), including the camera that captured the Thousand Oaks case, said, "Any time there is an incident reported on in the TV or in the newspapers, I get a lot of orders."[100] Nanny agencies and associations walked a tight line between respecting the parents' wishes for security, while demanding respect for worker privacy in turn, some making new policies for the nannies they represented to be notified of any cameras on site. Other local agencies said there was no way to know if cameras were in use, but that largely it was assumed. Others, even those in charge of protecting nannies, said that if the level of employers' suspicion was that great, a new childcare worker should be found.

These cases persisted over the next decade, wherein the media mined the images of mediated mothering to justify further surveillance, and childcare workers had to negotiate the new limits put on their freedoms at work. But the nanny cam was only written about as beneficial to parents—even if organized childcare workers' points of view were sometimes taken into consideration. The nanny cam was presumed neutral: it just captured what was there to be captured. For doing its job, it was routinely celebrated in the press. This included its role in the case of infant Laura Schwartz—whose parents, Jennifer and Bret Schwartz, were, by their own admission, anxious first-time parents, who had taken their search for adjunctive care seriously, interviewing nine applicants, talking to provided references—and then also conducting multiple background checks and hiring a private investigator. They also placed two nanny cams. The parents noticed that Laura was getting fussy upon being handed to the nanny.

They decided to finally check the cameras, and allegedly found two incidents of abuse that looked like they could cause shaken baby syndrome and one case of their employee biting their daughter. The Schwartzes declared to journalists, "The nanny-cam saved her life." Jennifer Thibeault, whose company, TBO Tech, sold hidden cameras, noted an uptick in sales following the Schwartz case's publicity. She also noted that parents were largely ashamed of using cameras this way, "but we assume that, if they're putting it in a child's room, they suspect the nanny. . . . All the parents we talk to want the cameras shipped in a hurry."[101] This is what the nanny cam was supposed to monitor: the idyllic normalcy of a quiet bedroom—in order to patrol it for signs of the lethal and pathological.

Yet the very illness that became synonymous with the camera—Shaken Baby Syndrome—became the site of medical contest just as nanny cams became a standard feature in employer's homes. Even the doctor credited with enshrining it in American pediatrics expressed regret over its status as a differential diagnosis after forty years of cases, declaring that, with proper imaging and case history management of the child in question, one would most certainly find a preexisting condition that led to the severity of the head trauma.[102] Nonetheless, the images and videos that circulate from nanny cams showing enraged caregivers are still used forensically and perpetuate the diagnosis in courts despite this serious medical revision.

Bringing in a mechanized replacement for the nanny at first might allay fears of contagion, only to replace them with fears of the other impurities stemming from techno-parenting. There is always someone or something from outside the family invited in, which brings that otherness with them. We might even say, paradoxically, that the extra-familial is intrinsic to the nuclear family, but tends to produce a persistent reckoning and a changing set of affects: relief, guilt, envy, fear, and suspicion among them.

From the 1990s onward, these technologies have become increasingly folded into mainstream consumer practices even as they have faced intense scrutiny for attempting to automate a mother's love and attention, much as the nanny, by mid-century, was understood to be the sign of a reduction of maternal investment and care. These technologies are recruited to blur that which counts as personal service in kin networks (love) and what might be waged work, who might do it, and under what circumstances. Even speculative technologies—such as AI homework monitors and robot nannies—might tell us that the panic around human otherness in child-rearing remains with us into our present. Here, following Lily Irani, both difference and dependence are rearticulated in equating the nanny with machine.

Yet "surrogate humans" have not—in childcare—by and large displaced childcare workers. Unlike voice assistants in our present, which remediate the domestic worker or the personal assistant, the nursemaid and nanny were already surrogates whose difference from a mother is constantly renegotiated as technologized family management and family care increasingly merge. The nanny panic, which is central to remediation of care in the bourgeois nuclear family, depends on conflating waged care with a form of negative transmission. Nevertheless, the promise of extending and augmenting parental nurture and protection has driven the marketing and development of much parenting technology since, which has grown to include monitoring tactics absorbed from, or associated with, more suppressive forms of surveillance. Many of these technologies encode the same class-based suspicions of their predecessors. Today, state-of-the-art parenting technologies are frequently designed to monitor not only children, but also those suspected of posing harm, making targets out of bystanders and importing state surveillance—inseparable, as Simone Browne has shown, from a history of racial formation and violence—into the home.[103] Surveilling children is part of parenting; contemporary parenting mores have intensified this basic imperative to watch, even as it is outsourced to care providers both paid and unwaged, to automated machines and their analog counterparts.

This was always the threat of the nanny: the possessiveness around maternal love, its demands, and the charge to preserve purity all are proxies for race and class. Nearly all families seeking to hire a nanny belong to the upper classes—nannying is almost always the most expensive form of childcare precisely because it promises greater attention to the child, and greater control for the parent as employer and worker themselves. Parents now pay to reverse the fungibility slope into automation by inviting in a human other with its promise of para-maternal presence and attention. It has only been in recent decades that psychologists have begun to attend to the difficult role handed to contemporary nannies by their employers and the impact of multiple caregivers—one waged, one not—on children. Psychoanalyst Susan Sheftel notes that we largely do not talk about the role and labor of the nanny as intrinsic to the reproduction of middle-class and upper-class families, despite these being the major population still seen in psychodynamic therapies.[104] Psychologist Daphne de Marneffe also attends to this care from the position of the mother at work, writing, "Turning the care of our newborn, baby, or small child over to another, non-familial person, someone we often have known only briefly, is a momentous emotional and psychological act, even if we pretend that it isn't."[105] What she leaves out—and what is left out from almost all psychological accounts, is what the work is emotionally and psychically to the person providing the care, who must negotiate both their and their employer's affects surrounding this highly charged form of labor.

Almost all nannies—still, in the twenty-first century—are women. Almost all come from a different socioeconomic class than their employers, and they may be of a different race (estimates vary on the racial breakdown of nannies—due to immigration status, under-the-table work, and other factors).[106] When parents want to return to work, feel they must, or need to in order to survive economically, those who use paid attention and adjunctive, diffuse care (as opposed to kinship networks) to cover their work schedules often enact some form of education, monitoring, and indeed surveillance on their new employees—especially those located within the home rather than beyond it.[107] Nannies and ad hoc babysitters may be people who parents trust enough to look after their children, but the widespread use of the nanny cam shows that trust to be partial. Surveilling the nanny is a displaced form of maternal presence and attention—watching the watcher of the baby instead of the baby itself.

What are we to make of maternal labor—a work figured as pure love or as rotten, wherein any form of assistance (a lessening of suffering for the assisted) is therefore a dilution, an intrusion, even if that form of assistance is also enshrined as a signifier of proper maternal care, whether in the form of a nanny or the device deployed in her stead? This double bind produces an affective, emotional, and material vicious cycle. Attempts to mitigate the bind only produce increased vectors of regulation (by experts) and surveillance (via technology). Or we can say that the affects attached to parenting have been exploited to *preserve the bind*: automation—here, of the nanny—is called into the home to replace or extend human forms of care and help, waged and unwaged, while simultaneously reifying mothering as a uniquely unmediated zone of reproductive labor. In this chapter, we have considered the nanny as doing the work of standing in for mother. The nanny's labor has been repressed to be remediated: media have largely been brought in to stand for the nanny. The rest of the book will look at this confluence of mother-as-medium and media-as-mother. Behind this, we will see, is the haunting of waged and unwaged childcare, the specter of nanny-as-medium as competing interference with mother-as-medium. Although "mother" is, under this set of normative understandings, the preferred term, she too has been put under extraordinary surveillance, one often first placed on the figure of the nanny.

2 OUT OF THE CRADLE

In 2016 the Snoo Smart Sleeper debuted to an incredulous American public. Disbelief presented in two ways: that the device might do what it claimed it did and that anyone would pay for it. The Snoo, modeled on a mid-century modern-inspired bassinet with thin wire-frame legs and a warm wood base, was marketed directly to expectant parents. It promised a salve if not salvation, taking care of one of the most notoriously difficult features of infant rearing: sleep.

The Snoo is a bassinet that is said to issue "constant calming womb-like motion and sound," rhythmically rocking a baby to sleep. Also described as an "extra pair of hands," the machine automates the tiring and tireless work of shushing and rocking the baby to sleep, reducing the maternal body to a womb and hands that tend to the prenatal and neonatal child. While these two innovations—white noise and automated rocking—have each been available across the twentieth century, the Snoo takes over the need for a parent to even deploy them. It is a fully automated system. Rather than a parent's watchful eye, the machine-bed itself has sense detection monitors responsive to non-pathologically fussy babies, joining the baby monitor to a sleep device. The Snoo falls on the side of what is called independent sleep—which is exactly what it sounds like (babies sleeping without a parent). A gentle, even warm, mother-like alternative to the "crying it out" method, the Snoo promises no tears, and independence through automation for infant and parents alike.

Invented by a married couple, celebrity pediatrician Dr. Harvey Karp and his business partner Nina Montee Karp, and designed by Yves Behar and Deb Roy of the MIT Media Lab, the device provides a simple promise to the working parent of today: if parents are better rested and can relax knowing their baby is safe, they will be better parents (and more successful employees). If the child is managed by a machine at night and caregiver by day, the family unit will cohere, with better results. As we will see, a rested family is a hallmark of what is understood to make a healthy family across the twentieth century and into the present. The only way a family with a young child might rest is if the child does. Accomplishing sleep is thus, since the second great industrial revolution, paramount for labor occurring both within and beyond the house.

The Snoo makes several fundamental assumptions about the family in which it sits: that families are conforming to the working day, that continuous infant sleep is paramount, and that they will turn to expensive technologies to achieve these aims. In short, these sleep solutions rely on what I call "neo-scientific parenting." How did a sleeping device like the Snoo, and its attendant parenting principle that good sleep is the basis for a strong family life and productive work life, come to be?

In its original form at the end of the nineteenth and early twentieth century, the practice of scientific mothering cast mother as a medium between the state and its experts and her child (see the introduction to this book). Scientific mothering, the belief that mothers are not instinctual and thus must be taught by scientists and experts, was employed across all the scenes and gestures of mothering and was particularly concerned with feeding, growth, and sleep.[1] Mothers were instructed to study the art of parenting as a science, and to track their infants, keeping intensive logs of their behaviors. The re-rise of scientific mothering, or neo-scientific mothering, shares in this: mothers once more are tasked with keeping records of "leaps" and sleeping, feeding, and defecation—often hosted through third-party proprietary apps marketed via the experts whose procedures they use.

Under the logics of early scientific mothering, mother was essential to family care, and her instincts to mother were in need of intense regulation. Often marginalized in histories of scientific mothering are the very tools that accompanied its regimes and allowed them to take hold. In this chapter, I will examine how mothers have turned to technological supplementation for a wide variety of parenting practices (what are now called parenting "styles"), from the Progressive era's scientific mothering to the neo-scientific parenting trends in late-stage capitalism. If in the previous chapter we looked at how the middle and upper classes sought to remediate the nanny for daytime care, in this chapter, we will look at how mother was reconfigured as a medium of care in the night, joined by a whole host of new adjunctive technologies. Simply put, the problem of infant sleep co-emerges with shifts in labor, both beyond the home and within it. No one much cared about the caregivers' nighttime (only their daytime productivity) until it's the wealthy who are doing the night's work.

A history of infants, caregivers, and sleep might begin with the bedtime rituals that caregivers have performed—lullabies, bathing—or the cultural scripts that accompany nighttime acts of social reproduction.[2] But what matters to the maternal media story told here, rather, is how the triad of crib, caregiver, and infant—the mediated assemblage of mother-baby—took on new cultural and social meanings across the twentieth century, when the crib became the designated proxy for the entire maternal environment. The crib (the word's earliest meanings are "manger" and "stall") proved to be a site saturated in rules, advice, and and technological augmentation.

From the turn of the twentieth century onward, the sleeping infant was increasingly put under intense observation, first in the laboratory called the home, and later in the clinic. Infant sleep was posed to caregivers as a dual problem: it was at once potentially dangerous—associated with smothering and cot death—and difficult to achieve. It needed to be a site of purity, or *non-mediation*, but also relied on devicing. How and where an infant slept took on new legal, pediatric, and scientific meanings as the normal place of sleep was redirected definitively to the crib, and bed sharing became pathologized as a parenting choice. These new meanings of infant sleep depended on a move from assuming that the mother (or caregiver) would sleep next to baby to prescribing that she should not. Put another way, sleep would no longer be procured through a largely unmediated relationship between maternal and infant body, sharing a single surface, and instead required that technologies step in to regulate—to mediate—mother and baby. We will see that infant sleep closely resembles what Roland Barthes calls a "fashion system"—where he describes a predictive turn and return in fashions (such as hemlines) based on what has just occurred. Highly mediated attempts at getting the infant to sleep are then followed immediately by new theories of unmediated care.

Once mediation became required for "good sleep," a whole set of new tools—with their own attendant parenting ideologies—were designed to keep babies safe while pacified, contained, and resting. The meanings attached to each of these tools have a history, and not one that centers on efficacy (or not only). Instead, these methods move in accordance with understandings about safety, individuation, and independence in childhood—and assumptions about how "natural" parenting either encourages regulation or increases unsafe sleep conditions. The aim of tools for infant sleep is, of course, to safeguard and control baby and, implicitly, to reshape the maternal medium.

This chapter turns to soporific and atmospheric media and what they tell us about how structures of time and labor shape scenes of care and maternal norms. From the first cribs thought to support the individuation of children through to the automation of that same autonomy in the bed of the Snoo, infant sleep has been used to regulate both the mother and her infant, often by reprising and remediating the mother via a proxy that removes mother herself from the side of her baby, supposedly conferring independence and fitness on children and preparing them to enter the labor market by conforming their bodies to its time signatures in infancy.

NEW WOMAN, ELECTRIC ARMS

From 1870 to 1890, the "new woman" emerged, less content with the overwhelming number of children women were expected to produce, and thus a challenge to the meanings of maternal labor.[3] What the role of the mother might be in safeguarding and impacting her children was dealt a decisive "body blow" by these changes in maternal status, as well as by the popular reception of Charles Darwin, whose theories were taken up by those who asserted notions of maternal impression over care as mattering for childhood outcomes. As we saw in the introduction, evolutionary theories were articulated in nascent debates about child development. If, as some argued, babies had inborn inheritance and potential, a baby's future was more or less determined.[4] A mother's role—as medium of birth and as medium of care—was newly contested, and the valuation of motherhood was reconfigured contradictorily. On the one hand, her child's nature was written by the time of its birth, and on the other, children's physical safety was paramount and needed to be ensured by its environment.

As historian Kelsey Henry argues, post-emancipation racialization became deeply embedded in these same schemas of child development and pediatrics, while white children became, in sociologist Viviana Zelizer's words, "priceless" and needed to be *protected*.[5] Put more simply, with these historical developments, the question of how much a mother should mother—how much labor she should take on in the family—was put under new, social pressure and was recognized as a social problem. One answer for reconciling this new era of maternity with this new era of white childhood came overwhelmingly from the parenting literature: mothers were to move against maternal instinct in order to become the proper medium of *care*. In one popular manual for children and their mothers, Baroness Bertha von Marenholtz-Bülow wrote that "the mother must no longer be content to meet her child's impulses with the maternal impulse—for impulse is blind."[6] Impulse is not considered instinct in this moment. An instinct is a way of knowing, while an impulse is an unthinking response. Something external to the mothering body and psyche must be called in to regulate that impulse, conform it to knowledge, and turn it to instinct.

The evidence for the shifts within parenting discourses is multiple, centered on new literal investment (in the form of mother's and widow's pensions) not only in mother but in technologies that would rearrange the stuff of mother and infant relations. These technologies largely had one aim: to regulate baby without proximity to mother. These new intimate architectures were invited into the home and thus asked to intervene on the mother and child dyad. Breastfeeding, the classic bellwether of the United States' relationship to mothering, was out of vogue and would come back only some fifty years later in the 1920s; formula took precedence and with it bottles that needed to be cleaned regularly.[7] The rise of high chairs and new modes of feeding older infants followed as well (see chapter 1), providing that added benefit of containing children. Prams, too, became used to distance mother from child or nurse from a charge, more a labor-saving device than a time-saving one, a way of being with baby while being with the self.

Sleep was a suddenly an essential site of intervention, driven by the array of social meanings sleeping arrangements had. Despite an early focus, replete with devicing, on regulating the caring body, even by the mid-nineteenth century manuals did not address the difficulty of getting baby to sleep. There was an explosion of parenting literature but, as historian Peter N. Stearns observes, infant sleep was "less contested and less orchestrated."[8] Stearns leaves open the question of whether sleep became more difficult or if managing the rhythms of family life did. Why scientists began reimagining familial sleep and provisioning families with devices so that they might put infants to sleep in new modes requires a social rather than biological answer. As we shall see, feeling rules around appropriate contact in sleep, distance, and both infant and maternal regulation dovetailed with the wholesale scientific reorganization of domestic spaces around child sleep. As scholars of US sleep patterns have noted, during the rapid rearrangements of daily life on the clock time of industrialization, sleep was becoming understood as essential and more elusive. For the whole family to be able to rest at the same time was of ever greater importance.[9] By the 1880s, sleep became understood as the foundation of good health, not just physically but mentally, not just for children but for adults. "Good," for the child, meant independent.

The turn to thinking of infant and child sleep as a problem to be solved concerned its material conditions: its spaces and smells, its positions. Handbooks and advice columns might address the dichotomies of good and poor sleep: hay versus a feather mattress, prone (face down) versus supinated sleep (face up), pillows versus coverings.[10] If, in the mid-1800s, whether infants slept alongside the carer's body or on their own was not heavily dictated, by the late 1800s women's magazines and child-rearing handbooks began to preach the importance of independent sleep. As scientific mothering gained steam and its hallmarks issued downstream in women's magazines—newly populated with advertisements for infant devices—co-sleeping became frowned upon for its necessary proximity between mother and child (independent sleep was, perhaps surprisingly, also said to curb masturbatory tendencies).[11] Those who continued to practice what was seen as an overly intimate arrangement became the intense site of focus for hygiene movements. This in turn was predicated on raced and classed associations with the practice, which then codified them such that they linger into the present.

Starting in 1890 and cresting through mid-century, there was a resurgence of mother as influential, as key site of nurturance, or, as I argue, the implicit theory of mother as medium was born from its older configurations as a theory of impression and impulse. As child-rearing manuals began to describe the child as "absorptive," questions of impression and impulse were reconfigured as attention and presence. The question remained: Were attention and presence *good* for infants, or, as many held out, did they corrupt them long-term? Where public commentators—including those from the nascent field of pediatrics and those aiming to use the family as a site of social reform—spoke, they issued new dicta about infant sleep. Nighttime care went from being centered on bedsharing, or holding, to beneficial separation.

As we saw in the previous chapter, placing a baby in a container for sleep became a proxy for waged (and unwaged) care providers. It served as a space of containment that let mother or other turn away without passing the child to another human.[12] Under logics that held that mother need not nurture particularly, subbing herself out for a rocking cradle that pacified infants functioned socially. But with scientific mothering, cradles, endlessly rocking, were rejected as a site of true independence. The repetitive motion was still understood as maternal, or feminine (as Wordsworth has it: "but with patient mind enforced / To acts of tenderness; and he had rocked / His cradle, as with a woman's gentle hand").[13] With scientific mothering, that very maternal gesture was pathologized. Rocking was thought to train children to crave uninterrupted attention, making them dependent on the motion like a drug—this held through the 1930s and was a foregone conclusion, as we will see, in the work of John B. Watson.

It is in this period that smothering went from a devastating but oft-occurring, naturalized phenomenon to an object of intense legal and domestic scrutiny. Smothering and co-sleeping now were understood as needing regulation, if not intervention. The maternal body, therefore, was understood as requiring its own regulation as it interfaced with baby. How to regulate this untutored, impulsive body was the task that Dr. Luther Emmett Holt assigned himself. Holt, a pediatrician and later a champion of Progressive-era eugenicist birth control practices, worked in New York City's Babies Hospital, becoming its first medical director in 1899. There, Holt observed not just an array of infants in need of care but how professionals—nurses—cared for them. Several effects followed. First, he became enchanted by the intensive note-taking nurses performed on clipboards that sat next to patients' cribs. This became an early medical chart. Holt was then in a position to amass that data and turn it into an account of child-rearing, sending the hospital home to the mother. This became his *The Care and Feeding of Children: A Catechism for the Use of Mothers and Children's Nurses* (1894). In it, he laid out three sections, two of which focused on feeding, both "natural" and "artificial" (Holt had been instrumental in quantifying the nutrients in breast milk in New York City). At the back of the book, in his fourth section on the "miscellany of the nursery," was a subsection on sleep. In it, Holt offered a prescriptive schedule (for an infant of six months of age, a sleep from 6 pm to 6 am save for one moment of nursing at 10 pm) and best practices—which included no bedsharing for fear of overlay *and* characterological defects, and no rocking.[14] About this Holt was serious: "By no means. It is a habit easily acquired, but hard to break, and a very useless and sometimes injurious one."[15] Holt was also against other forms of affection—like kissing—for characterological reasons, and equated rocking and other soothing gestures with that other maternal stand-in, the pacifier.[16] Throughout the manual, Holt expects the baby will sleep in a crib, in their own room—a vast change from the movable cradle and bedsharing that even one decade prior would have been commonplace.[17]

Independent sleep had a third connotation—associated not just with the mother and her labor, or her child's character, but with the child's physical health. Long before the sleeping infant became a site of scientific concern, its most terrifying outcome—smothering—first had to become disambiguated from natural if not accidental loss. As the early antecedents of scientific motherhood spread, those charged with care (the nurse if not the mother) and how they slept with baby came under intense surveillance as a site of increasingly preventable infant death. Tools that would allow for this separation now came from strange places, less the home or store than the hospital or clinic. New locations and new regulations urged an alternate set of arrangements for mothers and their babies, making these dyads anew.

If slumbering closeness was a physical threat, it was—along with kissing, as we saw in Holt's work—an emotional one, too. Yet a contradiction emerged. As middle-class mothers and their children were being separated to ensure independence, starting with sleep, technologists turned to automating the very drug we might call attention. Remediating rocking spurred US innovation in the nursery just as it was understood to belong to a bygone time. As we saw in chapter 1, remediations are often nostalgic, out of step with their scientific and social moment. Remediation is retailed, quite literally, as offering us in the present the stuff of the future. Here, we can see that technologists were out of step with pediatric science, constantly looking backwards to remake the present.

The first American attempt to introduce *mechanical* automation into mothering emerged twice: by removing the caregiver from the nursery, and then by replicating her soporific gestures, rendering infants independent of their caregiver if not independent of mimicked attention. These attempts to automate care occurred in parallel with early Progressive-era panics, both raced and classed, that attached to the figure of the nurse, and set in motion that long downturn in staffed homes that was more or less completed by the interwar period (see chapter 1). Rendering an infant independent had two desired effects: to make childcare labor less taxing and to make better child-citizens. Removing overnight care—the nurse—from the services in the nursery required new ways of tending the child in upper-class families to keep maternal labor static, not reduced in quantity but shifted in quality. While these panics were focused on the contagion of affects between nurse and child, additional shifts in conceptualizing a safe childhood were taking place at the same time. What a good configuration of mother and child, carer and charge, might look like was in flux, and the science of mothering was at the center of such change, medicalizing the infant in the nursery and mother in society.[18]

Until around the second decade of the twentieth century, sleep was assumed as the natural consequence of wakefulness.[19] As independent sleep became central to good parenting, a new paradox emerged: How to get an infant to sleep if neither caregiver nor mother is present? Presaging Taylorism, wherein each action of labor is broken into its constitutive gestures to be made more efficient, one answer came in the form of electric arms. Remediating perhaps the most iconic and physically taxing gesture of care, the rocking of a baby, the "Electric Cradle" removed the activity from human limbs (for even a non-mechanical cradle required something to set it into motion and keep it going), ceding it to electrical gears. Even though very few residences had electricity in them, the Electric Cradle made its triumphant appearance in the United States in 1898.[20] The futural gesture of automation offered dynamic life compared with the static crib, increasingly in vogue, while presenting itself as a neutral, underfeatured alternative to the alarmingly over-featured nanny (see chapter 1). Care was reduced to a gesture, one that could repeat ceaselessly.

Inside the home, early mechanical rocking cradles were quite simple and had but two electric components: a small light show and an automated rocker. The rocker function had two speeds: one to soothe and put a baby to sleep, the other a more vigorous setting to entertain the baby when awake. Thirty years later, in 1928, the first electric cradle patent was taken out by Herman H. Millard for a motorized cradle that simply featured automated rocking.[21] Across the rest of the twentieth century, small changes to the idea of an automatic, rocking cradle resulted in a rush of patents: 1947,[22] 1951,[23] 1972,[24] and 1988.[25] The promise remained the same: to automate maternal attention and separate it from maternal presence. Much like night lights were understood to confer a quality of maternal attention, raising the specter that they were training children to be needy and to desire more mother than was permissible, the cradle too carried this pathologizable media-as-mother logic. Unsurprisingly, it was quickly understood to be improper in the childrearing literature of the period.

The automated cradle communicates an infallible attention in rhythm. Depicted in figure 2.1, the cradle is supposed to be used while a mother works, most likely for herself (although it is not impossible that, given that

Figure 2.1
"The Electric Cradle," *Chicago Tribune*, February 13, 1898.

this scene is speculative, she is represented as doing paid piecework for others). Rather than placing the child with the nanny while she performs other domestic work, this middle-class woman can be a mother and housewife at once. In doing so, she is understood to remain the medium of care; she communicates presence even as she is occupied with other reproductive labor. The baby conforms to the rhythms of production, appeased by mechanical reproduction. The demands made by this gendered labor inflect and shape the maternal labor available to the baby: the baby must be out of the way, or at least sufficiently exhausted.

The mechanical cradle in the home was just one technical solution to the infant's sleep—and it often had complements in other containing devices used in the day. Baby jumpers were retailed and deployed to mix exercise and containment, designed to fatigue a baby and prepare it to rest. These inventions required no electricity, but ran on the motor of the baby itself. Here, the baby is ostensibly contained, soothed, and entertained so that mending, for example, might get done. The definition of feminized technologies and feminized labors converge in anything that can be set down and picked up again in relation to these very rhythms of child-rearing—knitting exchanged for baby, for example. In the arms of the electric cradle, a baby is exchanged for labor and the labor of mothering is supposedly obviated by automation.

INPUTS AND OUTPUTS

For commentators, if not caregivers, the problem of infant sleep inherited by the twentieth century was that of the characterological consequences of sleep patterns—and suffocation.

Other forms of crib innovation that promised to shift the duty of entertainment, health, or sleep from care to furniture, such as the window crib, appeared in the same period, and were used both in tenements and by the First Lady Eleanor Roosevelt. Whether by exposing infants to fresh air or allowing them to exercise, the environment itself was understood to contribute to good health necessary for baby getting to sleep and remaining asleep.[26]

Suggestions that a baby's sleep might have something to do with its character became a site of scientific study when programmatic scheduling for infants reached its height in the 1920s. Correspondingly, new infant sleep problems emerged from the tie and conflict between children's schedules and the productive schedules of their parents. Whether in newly aggressively demanding schooling or because domestic infrastructure made it difficult to separate children from parents, children's sleep was contested. Media themselves—those same dispensers of parental advice—began to be presented as cause for concern. Responsible for the overstimulation of children, media were understood to be the enemy of sleep (books for girls, radio for boys).[27] Over-attention—both mechanical (night lights) and from caregivers—and its resulting characterological consequences had become pitfalls for parents to avoid. Stimulation—from media as maternal proxy or from mother-as-medium—was now of paramount concern as scientific motherhood became further entrenched.

If some were worried over the longer-term effects of over-attention on babies themselves, scientific mothering was also supposedly of extreme convenience. Scientific motherhood allowed reproductive life to suture itself within productive life, or, as summed up by the famous dictum from behaviorist-turned-parenting-guru John Watson, "Make baby's time-table fit yours."[28]

As we saw in the preceding chapter, John B. Watson domesticated his lab-based conditioning practices. Watson, who writes warmly and indebtedly to Holt in his book *Psychological Care of Infant and Child*, noted that, thanks to Holt and others who wished to standardize parenting—as we might think of regulating and standardizing media—the cradle had been firmly banished, and now would only be seen, so he thought, "in a museum."

The baby, on Watson's understanding, thus becomes a vehicle for predictability, regularized like a train. We can also hear an echo of how the worker's gestures had been, in this period, reconceptualized in the factory. The idea of *training* infants for their job—to sleep, to eat—was born in the same period. It was an invention of necessity: in and among the soft propaganda for scientific motherhood was the caution that, with the decline of domestic service, children, where they could, must be trained, much as servants were (see chapter 1). Once trained, the longer they were asleep, the better—for their health, and also for their parents'.[29] For the youngest children, the most susceptible to training, the crib was, starting from Holt, the ideal space to set them down to grow independently.

Given that the nursery became a site of scientific attention in the early twentieth century, it makes sense that its chief marker—the crib, that static space of independent sleep—moved between the clinic and the home. In the nursery, the crib had been beta tested and was shown as encouraging independence (and was inattentive, compared to the rocking cradle that soothed). When the crib followed the infant into observation, it became part of the backdrop for infant development studies, much as it had informally been a cornerstone of proper development in the home. It was the necessary container for infant sleep studies, supportive furniture for pediatricians working to *standardize* the infant. In an effort to contain and standardize their subjects, pediatricians did as mothers do, placing the baby in its common, if not "natural," surrounds.[30]

Behaviorists continued to address parenting during and after World War II, but combined their psychological expertise with more complex devicing.

Watson's book had already disseminated behaviorist advice to the home. Now conditioning devices would follow, leaving the clinic for the nursery. In the late 1930s and 1940s, behaviorists addressed the disciplining of children via technical interventions into the work of child-rearing. In 1938, two psychologists, O. H. Mower and Molly Mower, addressed the common childhood problem of bedwetting by fabricating a device, eventually produced by Sears as the "Wee-Alert" bedwetting alarm, which, in is earliest versions would wake children up in the night at regular intervals to ensure that they could void; later, it claimed to alarm at the first contact with urine. As Deborah Blyth Dorshow writes of the device, "Compliance was difficult to achieve among children and their parents, relapses were common, the equipment was out of the price range of many eligible candidates, and when it failed to work properly, alarms could injure children."[31] Other children subjected to the alarms avoided being awoken by using various clever workarounds.[32] Despite these problems and the possibility of evasion, the use of alerts for children remained popular across the next five decades, in part replacing the nurse's overnight duties and, here, figured as cutting down on laundry. Such technologies did not address the problem *child*, but a particular issue such as bedwetting. Yet for the parent who wanted a total system of care, one alarm for one possible problem would not do. If an alarm combines the notion of presence and attention, it is too narrow. It merely intervenes in what is, then relays that data to sleeping parents. It is not a self-sufficient, closed circuit; it does not generate an environment.

Figure 2.2
"Wee-Alert" advertisement, Sears catalog, ca. 1950.

ATMOSPHERIC MOTHERING

In 1944, B. F. Skinner set about to combine all three features of the maternal and remediate early infant care entirely. Skinner sought to address the universalized problem of early child-rearing—that it is labor-intensive—and turned once more to furniture to remediate not just the nanny, or the domestic tasks associated with child-rearing like laundry, but mothering itself. Crucially, he set up all care as following from the space of sleep. He called his invention simply the "Air Crib" (frequently, and tellingly, mistermed the "Skinner box," which was for rats and pigeons and performed operant conditioning on them, equating the remediation of maternal labor with a nonhuman lab experiment). Skinner's crib was the first attempt to render a total system for infant care. It was ostensibly made for his second daughter, Deborah, but more accurately, for his wife Yvonne, who collaborated on the design even as the device was considered in the press largely as Skinner's alone.

As B. F. and Yvonne Skinner approached having a second child, Yvonne, "knowing what she was in for," experienced dread at "the drudgery of the first year,"[33] after the birth of their first daughter. B. F. Skinner purportedly thus rose to the challenge, reporting the following year in *Ladies' Home Journal* about the experience of designing the Air Crib in an article appropriately titled, "Baby in a Box," again an implicit denaturing of maternal care (a subsequent profile in *LIFE* magazine appeared in 1947). There Skinner writes, "In that brave new world which science is preparing for the housewife of the future, the young mother has apparently been forgotten. Almost nothing has been done to ease her lot by simplifying and improving the care of babies."[34] The Skinners attempted to do just that. They mapped out the daily schedule of this young mother—following behaviorist principles—and triaged what was deemed to be developmentally important. In response to what remained, they began "gadgeteering."[35] Several problems of child-rearing were addressed: the frequent laundry, self-amusement, cleanliness.[36] These jobs all fell to the mother. The Skinners reported astounding success with the invention: the baby was never sick and never cried (but remained communicative in other ways). These detailed benefits flowed from the crucial part: she was safe from smothering during sleep.

Skinner sought to make an atmospheric medium for childcare in an era where, as we saw in the introduction and first chapter, mother was being increasingly termed a "holding environment" or simply "environment" herself. The Air Crib, which was also sometimes jokingly called the "heir conditioner," also emerged at a time where time-saving devices were domesticated to save the work of housewifery. The dishwasher, for instance, was to save the time of doing dishes so that housewife was liberated to do her *other* work: mothering. The Air Crib was in part received as either miracle or abomination precisely because it was rather singular: it was a time-saving device addressed to mother *as parent* rather than as housewife. And in attempting to save time from parenting, it marked parenting as labor, rather than as unmediated love.

As B. F. Skinner understood it, the result for infants too was nearly miraculous: "It takes about one and one-half hours each day to feed, change, and otherwise care for the baby. . . . And after all, when unnecessary chores have been eliminated, taking care of a baby is fun."[37] The Skinners' personal pediatrician thought the device was wonderful—and *Ladies' Home Journal* had a number of unnamed psychologists and doctors evaluate Deborah Skinner, all of whom concluded she was very healthy. (Perhaps her role in the Air Crib gave on to pernicious rumors that she had gone mad, as Deborah Skinner writes, "The early rumors were simple, unembellished: I had gone crazy, sued my father, committed suicide.")[38] Pediatricians and their patients began to want the device for themselves, writing to Skinner at Harvard to ask for the device's specifications.[39] In parallel, some indeed were scandalized, including some of the Skinners' intimates. These two reactions—joy at the automation of the maternal function and subsequent partial liberation from the drudgery and sleeplessness of child-rearing and the indignant upset that anyone could abide that same liberation—help identify the conflicted feelings around automating mothering emerging at mid-century. They were articulated in telling contrast to the pervasive and self-evident welcome given "time-saving" devices that automated the labor of domestic chores. One form of devicing was acceptable, the other contested.

While some worried about the impact on the baby's temperament, much like earlier devices were thought to erode character (that being catered to

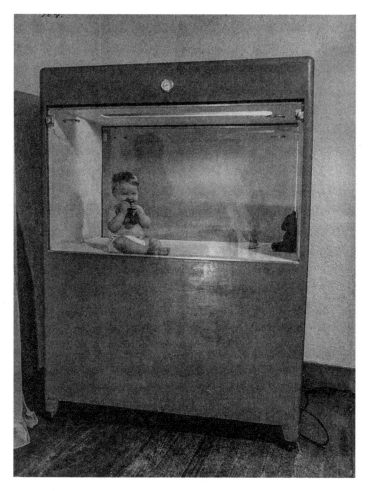

Figure 2.3
John Gray Jr. in an Air Crib, *LIFE* magazine. Photograph by Bernard Hoffman, 1947.

would make the baby temperamentally "soft" or too rigid), most of the critiques centered on two interrelated fears: that any form of artificial care was essentially against nature, and that going against nature had consequences—something would go wrong with the mother and her child. Skinner retorted that what was artificial in his crib occurred naturally elsewhere, and as historian Karin Calvert shows, the very turn to cribs and away from the cradle in the first place was initially a response to removing the human nanny and replacing her functions with furniture.[40] Skinner wrote that "some of the friends and acquaintances who had heard about what we proposed to do were rather shocked. Mechanical dish-washers, garbage disposers, air cleaners, and other laborsaving devices were all very fine, but a mechanical baby tender—that was carrying science too far!"[41] As David Snyder writes of this moment, "A burgeoning middle class grew out of a cultural context that privileged leisure time over work . . . soon nearly every consumer product was 'quick and easy' or 'timesaving.'"[42] Time-saving was not to mix with *mothering*, but to allow for more of it. Where it did, mother would be shamed: mothering was not to be limited to make room for leisure.

The last critique centered on deprivation: any automation of maternal care would leave the baby "robbed of affection and mother love."[43] This was cast as a matter of nature that evades innovation tautologically. To reject the disciplining of the baby via device, there must first be a fantasy that a baby is not yet disciplined by the comings and goings of the family, by familial labor and its resources, by parental need in response to these other demands. Skinner wanted his baby to sleep through the night, sleep longer, and sleep better so that he could, and so could his wife.

As in previous moments of infant sleep innovation, shifts in both productive and reproductive labor produced the contours of technological imagination. This was, after all, a moment where white women of Skinner's class were increasingly entering the workforce. The Air Crib put babies on the schedule of waged labor long before they were productive, disciplining them to productive time rather than that of reproduction. Deborah's parents were able to control her schedule by warming the crib before she'd wake, so that she'd persist in sleeping until they were ready for her to rise. The baby slept while mother was sleeping, rather than the other way around. Skinner saw this as a victory. As he told a number of "law wives" at Harvard, children are masters of operant conditioning; when they cry, parents pick them up.[44] Skinner saw this as a way, more or less gently, to turn the tables, for the apparent benefit of the whole family.

Although Skinner makes the sound point that a mother who can manage a baby well will be more loving than a resentful and exhausted mother, management is an ambivalent aspiration that represses and elides other concerns. After all, conceptions of "nature" and "natural" attach themselves to mothering, especially under essentialist conceptions of such; automation maps to the twentieth-century story surrounding masculinized labor. Skinner's reception redounds to the popular axiom: you can't automate love, or its tokens—the triad of attention, presence, and environment central to the mother medium—without provoking shame and revealing reproductive labor as such. You can smuggle all the work of mothering back in, and all the fear of the artificial, under this collapse of love and labor: intensive mothering then becomes the most loving. On the other hand, many parents now *do* discipline their children in the paradigm of neo-scientific care—conforming them to micro-markers of development, whether via sleep training, introducing foods in particular sequences, or restricting feeding to particular times. The difference: the Skinners had a machine to do it, taking over for mother.

If Skinner's reputation as a "behavioral modifier" or a "zapper" damned the reception of the Air Crib, it was partially because the assumption was that the infant was being modified, reduced to a little experiment. In reality, Skinner argued that the crib was *for* the baby, a better solution than Holt's and Watson's standard crib in its own room—or as Skinner had it, a "little jail."[45] For Skinner, the entire family could undergo operant conditioning, and its schedule and its gestures of care were all on notice for revision. The machine became a mediating instrument between parental labor and child; mothers modified the machines (see figure 2.4) rather than only directly attending to their babies. Air Cribs were also movable rather than necessarily isolated and static. Although babies slept independently, the Air Crib also was their place of play, and accordingly could be pushed by a parent to any room.

Nonetheless, the Skinners were able to market their own device; a few companies produced Air Cribs and hundreds of children were raised in them between 1957 and 1967 (including Skinner's own granddaughter, Justine).[46] In the late 1940s, Skinner's correspondence reveals a dogged attempt to collaborate with pediatricians, the main interface for parents who might purchase this form of care and the freedom it represented, to get the "baby tender" produced.[47] Parents, too, wrote to Skinner directly, hoping to build a crib themselves after seeing it in the likes of *LIFE*, the *Indianapolis News*, and

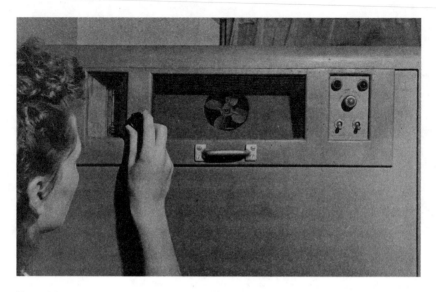

Figure 2.4
A mother adjusts the controls of her Air Crib. Courtesy of The Countway Archives, Harvard University.

the *Boston Globe*. By February 1948, Skinner estimated that over a hundred of his correspondents had made Air Cribs at home using his mailed-out specifications; Robert Topper, a psychologist at Emory, was helping manufacture them for parents in the South; and one company had produced several for the wealthy in Los Angeles.[48] He happily reported that one could be purchased for a "reasonable price" this way—of about $150 (nearly $2,000 today). Yet to Skinner's chagrin, he could not reconcile those who wanted the Crib and a mass market for automated mothering. These "Skinner children" (the inventor of the device displacing their patronymics) were treated as suspicious, wrong, off, or "freakish," despite being healthy, or alive at all.[49]

Dr. Spock—in the mind of the American public—Skinner was not. Despite the positive reception of the Air Crib in some medical corners, in 1976 Skinner was dismayed to watch a documentary with him as titular subject (Lawrence Olivier narrated) that called the Air Crib an incident from which he never recovered.[50] This same fate befell earlier pediatricians who attempted to bring mechanics into pediatrics for ease: to combine the mother of flesh with a *machine* was to redouble mother as medium, and it was met with panic.[51] Those who remained interested in the Air Crib were often other child development researchers, who hoped to use it to measure child activity, much like Gesell had used earlier cribs as a method of containment and standardization in his own clinic.[52] The clinic, that is, remained the dominant location for the development of mother media and the only place where the perceived coldness of its technology was understood as proper to its artificial environment.

SAFE SLEEP

Skinner's model of mothering was one that *desired* regulation, automating care itself. By mid-century, there was a robust literature on the rise of such regulation, and its corollary, "feedback." This was a hallmark of cybernetic thinking. When actual parenting met theories of conditioning, or regulation and feedback, the dangers of being misattuned were still high, and redounded to child death. Air Cribs have a theory of smothering built in; this was not the only theory and it accounts for the change in home use by those who adopted the device. The fear that was used to escort the practicality and scientificity of the machine—smothering—was not just a fear in 1944 when Skinner got to work on the Air Crib, but one ever increasing across these twenty years of the Air Crib's reception. We might think of the manufacturing date of 1944 as just before a revolution in mind for American parents: the release of Dr. Spock's *The Common Sense Book of Baby and Child Care* (1946). The volume—and Spock parenting generally—is often remarked on as ushering in a new era with its opening sentences: "Trust yourself. You know more than you think you do."[53] As we have seen, mid-century visions of mothering understood mother as all impulse, no instinct, and where instincts were credited, they were suspect. This pair of sentences, and the rest of Spock's manual that followed, ostensibly conveyed the polite but assured end of scientific motherhood. It might seem then that Skinner's machine was accidentally but quickly woefully out of step with mainstream American parenting practices and would have done better when Watson's behavioral parenting reigned.

Another, quieter change was taking place in pediatric medicine. Spock's rejection of scientific motherhood while continuing to produce manuals for parenting is an instance of legerdemain: he told his mother-students to trust themselves, while also telling them how to do so—the same paradigm with different sentiments expounded by scientific mothering. If only the affects surrounding parenting manuals changed with Spock rather than their intent, aspects of the content changed as well. About infant sleep, there was one single change: a new, widespread (and in practice calamitous) commandment to place infants in the prone position so that they'd sleep better and longer. This echoed the idea of mothers trusting themselves and what they witnessed: babies, the manual argued, sleep better on their stomachs. Rather than turning them over, leaving them be became, in Spock's hands, an instance of common sense. Spock argued simply that pronated sleep would bring relief and rest to the exhausted parent.

At the tail end of Skinner's production of the Air Crib, smothering, or what we now mostly call Sudden Infant Death Syndrome, had become re-identified as a signal worry of parenting in the 1960s. Air Cribs in their actual usage—in the homes where they wound up—revealed yet another change in infant sleep. Skinner collected surveys and testimonies of the children who were in his Air Cribs; many families sent back photographs. Arranged chronologically, they show that over time, families began to use the crib as Skinner intended: to keep a baby warm and thus free from needing anything in the crib that might harm them. They tell a story of subtraction. Even though the Air Crib was always, at its heart, a form of atmospheric media, one that was designed to keep a baby completely healthy, safe, and comfortable, parents must have wanted to add something soft to the crib initially, to make what they perceived as a sterile environment more cozy, more appealing. But across the dozens of photos Skinner received of these happy, healthy-seeming children, we can see fewer and fewer toys, clothes, swaddles, and pillows dotting the interiors of the crib. Eventually, babies are, as Skinner intended, stripped down to their diapers, kept warm by the Air Crib, and left to suck at their bottles.

What accounted for the change? Parents had always had the opportunity to use the machine as Skinner intended, but only started to do so *fully* in this period. As smothering worries became more precise in attributing suffocation to pillows and the like, the Air Crib functioned as the atmospheric and environmental medium that Skinner had wanted it to. It turned out that Skinner's insistence on fans and heaters within the crib, and the abandonment of all loose cloth, was deeply prescient of contemporary understandings of safe sleep (which would not become widespread in the United States for two decades). This is not merely a story about medical progress, wherein pediatricians learned more and more about what caused SIDS and thus made new infant sleep recommendations based off empirics. Instead, the rise in attention to infant smothering correlated to new sleep guidelines—those that supported letting infants sleep in the prone position—and they were deadly. Some public health scholars have gone so far as to argue that it is an instance of common sense: Spock's manual, which popularized prone sleep, alone contributed to rising infant mortality rates in this moment—and is estimated to have led to 60,000 infant deaths worldwide.[54] Yet Skinner never got credit for creating a crib-as-container that might protect families from this tragedy while addressing the double bind of productive and reproductive time. His operant conditioning of the family was too cold—too much blatant mediation for mediated mothering. Common sense was warm, and maternal warmth had passed from bad impulse to good instinct.

There is a simple reason why Skinner first made the specifications of the Air Crib for his own family: it is incredibly difficult to engage in medical research on children, let alone infants. To turn to one's own family as laboratory is very much in keeping with the history of child development research through this period. At the same time, as we have seen and will continue to, other pediatric researchers turned to groups that were likely to be available research subjects: children brought into clinics by anxious parents, babies in prison, and orphans. These children had departed the total institution of their families for other closed systems and, once there, became contrasted with the control group, somewhere far away, at home, in the first total institution of the nuclear family.

Spock was not alone in paying closer attention to maternal instinct in the name of science. And it is no wonder then that the developments of media to perform acts typically thought of as the work of a maternal regulatory system first look to other cognates for inspiration. The monkey, that other experimental animal, was often a chance site of inspiration for domestic technologies that now seem commonplace. Whereas Skinner had sought to make a near-total atmospheric system to remediate mothering, other child development psychologists moved toward mediated mothering, hoping to capture the entirety of the maternal environment via a single tool that could represent all her presence and care in one gesture.

THE SOUND OF MOTHER

This was what Lee Salk, the brother of Jonas Salk (the inventor of the polio vaccine), set out to do with what would become known as the Sonotone Securitone. Salk, a child psychologist who began his long career investigating the technical problems of mother-infant relations, was inspired by a chance visit to New York's Central Park Zoo in the late 1950s. There, in the Ape House, he saw rhesus monkeys carrying their babies over their hearts, and not just one time. Every single one of the mothers carried their babies uniformly. In keeping with moving from monkeys to human babies, as was in vogue in psychological research in this moment (as in the experiments of Harry Harlow; see chapter 3), Salk soon realized, in looking at parents he encountered, that human mothers tended to do the same. He formally tested his hypothesis on some 271 mothers, both left- and right-handed. Overwhelmingly, they carried their babies on the left side and over their chests despite side dominance. Upon presenting his findings to colleagues in Edinburgh, one psychologist with expertise in hypnotism ventured that the maternal body was indeed quieting baby with its rhythmic pulsing.[55] Of course, Salk's research was immediately understood in the Cold War paradigm of mental contagion and brainwashing, as well. On this understanding, the heartbeat recording became a sonic medium over which to panic, even as its presence was soothing.[56]

Salk then took his hypothesis one step further and tested the impact of the heartbeat on infants by broadcasting a recording over the speaker system of the New York Foundling Hospital. The hospital had been founded in 1869 in response to earlier instances of infanticide and child abandonment, just as smothering was being codified as nonaccidental. While the hospital was increasingly focusing on child welfare and less on medical services, Salk still was able to use a number of infant orphans (between sixteen and thirty-seven months) as his experimental subjects. The crucial criterion was that they were lacking access to their mother, and thus her heartbeat. Using the sound system, Salk tested the infant response to metronomes, lullabies, and his recordings of the maternal heart, as well as keeping the ward silent. His findings: the heartbeat was vastly more likely to yield calm infants, who also gained more weight more quickly. No sound *at all* was preferable to a metronome or to lullabies, that other technology of sleep. The Foundling Hospital, having been used as the staging ground for the experiment, resolved to make the heartbeat machine a permanent feature of its care.

In the staging of the Foundling Hospital, automating the maternal made a kind of sense: nurses and social workers were already doing the work of standing in for mother. A quiet hum, a presence of the maternal without mother being there had a salutatory effect. But when it came time for Salk to expand his research, he chose to make a home version for the domestic market, the Securitone Heartbeat Comforter. Securitone, a large white box meant to be set on a bedside table near a crib, was manufactured by Sonotone, a corporation most famous for its production of hearing aids. In advertisements for the device, parents were urged to bring the hospital-grade device into the nursery. Despite the intervening decades since Spock's revelation that mothers might trust themselves, techno-parenting still made use of the hallmarks of scientific motherhood, and appealed to parents to make recourse to doctor who knows best, even if it is to remediate mother.[57] Yet the advertisements actively sought to appeal to parents as a *natural* prosthetic extension of mother—a contradiction in terms. In the same moment where Skinner fought accusations of monstrous experimentation on children for automating their cribs, Securitone offered science without connotations of experiment per se, despite having first been tested on orphans in a hospital.

Salk was just one pediatric researcher who engaged in remediating the mother's body—here, the sound of the heart as it is heard in the womb—as early childhood care. In hospitals, new incubators were being tested on premature babies, including by Dr. Mary Neal, who constructed an atypical postpartum device: a swinging bassinet. Used on premature babies, Neal's device was meant to remediate that other maternal gesture of rhythmic rocking.[58] Neal's research reached the public through *Rock-A-Bye-Baby*, a *LIFE* popular science film on maternal deprivation, which made its debut at the 1970 White House Conference on Children. As we will see in the next two chapters, maternal deprivation research was a centerpiece of pediatric psychiatry and psychoanalysis in the post-war era. Here we can see that the absence of mother was crucial to the formulation of media-*as*-mother; researchers were consumed with diagnosing the impact of her absence and pursued anything that might attenuate it. Sent from the clinic to the home, where mother was assumed to be, these new technologies that presenced the mother in her absence helped escort media regimes of metonymic replacement. A technologized heartbeat, a movement might represence mother, allowing her to be elsewhere. As we will see in chapter five, by the 1970 White House Conference on Children, this logic had departed the confines of research, or the bounds of the nursery and its nurse. Media-as-mother then took hold in the living room, centered on television.

The logics of maternal substitution and proxy were being researched by Salk and Neal just as notions of kin and biological belonging were challenged once more. What constituted a family was once more open for discussion—by lesbian mothers, by advances in reproductive medicine, by shifts in notions of adoption—and artificial mothering became a site of intense investigation. Salk, for instance, did not see his work in technological fabrication as separate from these changes to family making and family life. After the great success of his device, and in an era where anti–artificial mothering campaigns became even more extreme, Salk weighed in on what makes a family in books like *What Every Child Would Like His Parents to Know*; held forth in his *McCall's* magazine column, "You and Your Family," for some twenty years; and routinely addressed the nation on the TV shows *Good Morning America*, and *Today*, guiding the nation through changing family norms and social panics. Salk too was called in to testify at the hearing for custody over Baby M, one of the first US cases in surrogacy law (where he sided with the biological parents—arguing that the surrogate had not yet begun to *mother*, and was only a surrogate uterine *environment*). Salk's understanding of the proxy environment—that it was reproducible, automatable—no doubt inflected his decision.

NEO-SCIENTIFIC MOTHERING

Within the middle-class family, the maternal medium—and her tools for care—were subject to continued refinement, as they also served as the grounds for bitter disagreement. For some, starting in the 1960s, the very devices that had stood for decades—the cribs, playpens, and high chairs—fell out of favor in many households. Rejecting hard furniture for something less structured (a sling or a mother's arms) and more focused on "skin-to-skin," the casual definition of unmediated mothering, a new parenting war presaged the rise of intensive parenting that would become a hallmark of the 1990s.

Those who used the old mode of contain and entertain were marked in opposition: they were suddenly practicing *detachment parenting*. Attachment parenting, an intensification of Dr. Spock's philosophy, came into its own in 1975 through Jean Liedloff's theorization of the "continuum concept," via her ethnographic work with the indigenous Ye'kuana in Venezuela. Liedloff claimed that, compared with tech-free Ye'kuana parenting, modern American technologized parenting made for less happy, less secure babies (for more on both attachment theory and attachment parenting, see chapters 3 and 4). The continuum concept does away with augmented motherhood, eschewing strollers, cribs, and bottles (especially with formula in them). The family bed and co-sleeping are argued for specifically as a "natural" way of sleeping that furthers bonding, needs no special equipment, and allows for mutual regulation of parents and child.[59] Much as those proponents of individualistic infant sleep derided these arrangements at the turn of the previous century in pejorative terms—"animalistic," unsafe, or unhygienic—we might read this deployment of Ye'kuana scripts for sleep as a plea for the return of an unmediated mothering—one that perhaps never existed for this class of mothers in the past. The nostalgia guiding this plea was not only a colonial fantasy of the untouched or pure sleep, but part of a larger turn in parenthood. A return to breastfeeding and "natural" and home birth practices in the same moment reinscribed a norm of "noninterference" and purity in the maternal bond, the legacy of which continues today in the adage "breast is best."[60]

This parenting style and philosophy becomes known as attachment parenting in the 1980s when the pediatrician William Sears and nurse Martha Sears began publishing mainstream parenting literature and advice on "immersion mothering," signaling devotion to childrearing while carrying with it the history of ethnographic participation in a culture not one's own. The pediatricians switched the coinage from immersion to "attachment parenting" to change the valence of their approach from what was most popularly connoted—drowning—to something positive and scientific, that is, attachment *theory*. And yet nostalgia, just as we saw in chapter one, pervades the recommendations, and Sears and Sears deride the stuff of everyday parenting life, especially the down-market solutions for containment, pacification, and play. Feminist theorist Cressida Heyes argues that, alongside the voyeuristic and appropriative reading of Ye'kuana family life, we can also read this turn in attachment parenting as reactive to white mothers' increase in their work and as "nostalgic for a family-wage era before austerity, and disdainful of a mother's independence as someone with her own private life apart from child-rearing and her own public life of work or citizenship."[61] If the introduction of certain tools was under the sign of nostalgia for an even earlier era—one of servants—then the removal of them could be understood as nostalgic for what followed, the era of the housewife, which had now also passed. The crib then figured as the appliance of the cold mother moving quickly for her own rest; bedsharing, now called co-sleeping, was the sign of the devoted mother. If, as we saw, the rocking cradle of the turn of the last century was a nostalgic technology, already out of sync with its pediatric moment, so too was the mother co-sleeping with her infant in the 1970s and '80s.

While co-sleeping and skin-to-skin contact had once been, under the sign of scientific mothering, the hallmark of "uneducated" parent, under the Sears regime they were reintroduced to middle-class American parenting cultures as both a duty and a luxury. Attachment parenting was a form of immediate mothering—as in parenting without media let alone a general recourse to stuff. Whereas the crib was a solution of the domestic, its years of development in the clinic ensured its status as both cold and scientific and the sign of proxy mothering. The mother medium was not to be replaced despite the logic of fungibility intrinsic to it. Rejecting Winnicott's decades-earlier championing of the transitional object as just that—something that helped children move smoothly from mother-baby to independence (see the introduction), Dr. Sears wrote damningly: "The child who is often left by himself in swings, cribs, and playpens is at risk for developing shallow interpersonal relationships and becoming increasingly unfulfilled by a materialistic world. . . . Teddy bears and baby bottles have helped us raise a generation of people attached primarily to material things."[62] The crib is redescribed as a jail.[63] Here, attachment parenting clearly disambiguates qualities of presence. The maternal body, rather than her material proxies, is necessarily mandated as attentive. Furniture, lights, and sounds were understood as proxies, and not sufficient. Indeed, under this conception, parents who had recourse to these proxies were told they did harm. This argument recalls the late nineteenth-century conflation of material objects with (over) attention. It also finds its echoes in surprising contemporaneous theories of child development, like psychoanalyst Wilfred Bion's notion that when infants cannot tolerate frustration, they turn to the world of material objects.

Arguments for more mediated mothering always coincide with arguments for less mediation. For every Skinner a Spock, for every Salk a Liedloff. As we have seen, each turn of the wheel rearticulates mothers, media, care, and political economy. This pattern followed into the late twentieth century, when attachment parenting was by no means the dominant form of parenting. Neo-behaviorist models instead were championed, and parenting-with-objects remained central even as it changed. If Sears and other attachment parenting literatures eschewed furniture, by 1985 a new parenting guru took them as central: Dr. Richard Ferber relied on them for his method of sleep training. Sleep training, and particularly the notion that children will "cry it out" or self-soothe, had been offered to US parents since 1894, when Dr. Luther Emmett Holt penned *The Care and Feeding of Children*. In 1985, using the "cry it out" method became known as "Ferberizing" infants. Dr. Ferber encouraged parents to slowly help infants build up a tolerance for independent sleep, culminating in them learning to self-soothe in the final phase. Despite the supposed revolution of Spock at mid-century, just one generation later, many parents returned to the two choices offered them at the turn of the century: co-sleep or have baby cry it out in the crib.

Just as biological reproduction, including an opening-up of who might be called mother, was being legislated and recodified, and care by mother was being intensified despite an increase in white women's work outside the home, this set of pressures put on a new generation of white middle- and upper-class mothers—the very same housewives to whom these devices have been historically addressed—made pursuing the techno-augmentation of mothering even more complicated and divisive. This moment did not eradicate attempts to automate *domestic* labor nor, despite media panics, did it do much to keep media out of the hands, eyes, and ears of children (the 1970s were, after all, the golden age of educational television, as we will see in chapter 5).

Nevertheless, the 1970s and 1980s did not see energies around personal computing and domestic relief for housewives converge at, say, an updated version of Skinner's automated crib (see chapter 6). Instead, the very basic devices that mothers (and other parents and guardians) used to take care of children saw no development, even as amateurs and hobbyists experimented with meeting their own needs through innovation and hacking, like one early Apple I user who programmed his computer to rock his baby to bed.[64] Generally, parents in the 1970s and 1980s used essentially the same devices that their parents had—or used no devices at all.

This was in part due to a divergence in parenting styles, wherein the basic crib was a tool for Ferberization and no crib was needed for more attachment-inflected parenting styles. The very place a baby slept was now indicative of a whole host of social and political choices. We can also note that the demands for synchronized familial sleep changed, tied to deindustrialization and the decline of industrial relations from which the conservative family movements benefited, which were strenuously preserved in the 1970s.[65] But the stasis in crib development occurred for policy reasons as well. Just after production on the limited run of Skinner Air Cribs stopped, in the 1970s and into the 1980s, several forms of contest converged at the level of policy and discourse. Pediatric standards for cribs were codified in the United States in 1973 in response to rising SIDS rates, ending easy experimentation with the crib form, either aesthetic or scientific (where experimentation did happen, it sometimes ended in recall due to child fatalities and still does, as evidenced by recent product recalls of vibrating, swinging, and folding bassinets). This stasis lasted through the "Back to Sleep" campaign of the 1990s—which drastically reduced SIDS.

THE MOTHER OF WORK

In the contemporary moment, crib tinkering has once again successfully captured a market. Whereas scientific mothering relied on earlier, standard technologies to achieve its aims, now, the present icon of mediated mothering—the Snoo, with which we began this chapter—goes one further: it not only coaxes carers toward neo-scientific mothering; it automates it. The Snoo, much like its likely neighbor, the sunrise lamp (most popularly fabricated by Hatch), promises to sleep-train a baby with few to no tears, safely, from birth. These soporific technologies promise first to control and then guide children toward "better" (read: longer), sounder sleep conducive to parental well-being. The underlying assumption is that these parents do not have care or kin networks to assist them in early parenthood. Unlike many earlier attempts to automate infant sleep, stigma is not what slows the production and uptake of this artificial environment—price is. The Snoo costs roughly $1,700 and works only for the first four to six months of a baby's life (the company eventually began the practice of renting them out as well, halving the cost). The machine pairs with an app, which alerts sleeping parents when the baby needs human attention; otherwise, the machine and the baby can be left to each other's devices. Nonetheless, if a simpler, mechanical bassinet costs under $150, a Snoo for more than ten times that amount is not a device within reach of parents who, were it shown to be effective, might need it most—those who are without maternity and paternity leave protections, for whom sleeping is essential to the return to work.[66] The Snoo, along with all sleep literature, presumes that the working day occurs *in the day*, conforming the baby to this specific vision of the rested baby and wakeful parent irrespective of the nature of labor.

For some who might be able to afford it outright, like the employees of "Big Tech," start-up, and media companies, employers are renting the machine for parents to return parent-employees to productivity as soon as possible. Here, the company aligns with neo-scientific mothering, and together they follow parents into their home to guide them back to work. As Susanna Rosenbaum writes in *Domestic Economies*, "From a middle-class (primarily white) vantage point, *proper* motherhood is exclusively concerned with children. . . . [This] understanding clash[es] with an analogous emphasis on self-realization through paid employment, and . . . collide[s] with the urgent need to earn an income as well as with the ever-expanding reach of the workplace. The fact that neither the demands of the home nor the exigencies of work have abated creates an increasingly unmanageable situation."[67] The paradoxical demands of neo-scientific mothering and "achievement feminism" collide. Work and home place women, as feminist scholars have argued for more than half a century, in a double bind. Mothers, as we have seen, are continually articulated as central to their child yet a threat to same, in need of regulation. Better sleep is figured here as a mother's way to achieve that self-realization doubly.[68] Now, being rested enough to produce, reproduce, and produce again is the ideal foundation of the working mother.

Yet the achievement of rest is not, for some, enough. The drive to know rest, in its intimate fluctuations, to chart it, quantify it, drives some parents not just to trackers (an old paper practice turned digital), but to devices that promise to track *and* alert.[69] It drives the extension of mediation of childhood into the night. Parental vigilance in sleep may be impossible, but prosthetic watchers resolve the problem by distributing 24/7 care across human and machine. Whereas the light sleep of new mothers and wet nurses is a long-documented phenomenon, devices like the Snoo, as well as stand-alone biometric sensors like the Owlet Smart Sock, introduced in 2007, aim to comfort parents while assuring sleep safety. They redress the gap between the interior and exterior, prediction and visible symptom, obviating the need for a parent to hover and watch their babies breathe. These devices, most commonly a smart camera placed over the crib or in a piece of clothing that functions as a pulse oximeter, claim that they will notify you if your baby is losing oxygen—preventing tragedy before it can occur. The device's job is twofold: to help parents monitor the biometrics of their child and to reduce worry by doing so. But the devices often do more harm than good in frequently transmitting false positives; and pulse oximeters work unevenly, with decreased accuracy when reading melanated skin tones. The result: "terrorized" new parents, who shift fervently between their infant and their devices, and who report a higher incidence of depression and sleeplessness. The false positives even clog pediatric emergency services—with parents begging for attention to infants they assume have lost oxygen, when in reality a sensor has malfunctioned or slipped. Whether through using managed sleep to aid in productive labor, or automated safety to monitor child, these new devices remain in keeping with Ruth Schwartz Cowan's adage about time-saving devices in the domestic home: they paradoxically produce "more work for mother," a function that now includes co-parents and other carers, and the unwaged labor of worrying.[70]

This is the domesticated sleep research laboratory that we call a nursery (even as, for adults, many now preach that the inclusion of technology in the bedroom is the enemy of good sleep). The infant's sleep laboratory deploys soporific, atmospheric media for care, and reemerged in relation to a neo-scientific motherhood, lack of universal parental leave, and the demand that parents be ready to work when infants are still at their most sleepless. But technology and architectural change have long played a role in sleep. As much as we might now be commanded to practice sleep hygiene, sleep is ever more technologized, and not only for the wealthy purchasing rest via expensive devices (or via employing nighttime childcare, or both).

Making babies predictable and schedulable is supposed to make them easier to *manage* via the magic of protocols and technologies. Recalling early forms of scientific mothering, a neo-scienticity has again found its place in the nurseries of the wealthy and the middle-class, bringing with it white noise (in use long before and long after Securitone's heyday), sunrise lamps, and automated cribs. With new furniture have come superficially new forms of digitally documenting child behavior, pediatric apps that disseminate advice and tools for scheduling and regulating the infant. But neo-scienticity features a serious revision to Watson's protocols: attention is no longer pathological, and proxy attention is called in to serve the infant in their sleep. These innovations of neo-scienticity are now sold—or rented—down-market. Now all are supposedly able to secure a womb remediation that encompasses attention.

As Haytham El Wardany writes, "Just as sleep hands itself to work, so work gives itself to sleep, each proceeding toward the other like head meeting pillow. Because work needs sleep."[71] Sleep may be the cousin of death, but it's the mother of work. Shifts in American sleep—and sleep research—are deeply tied to shifts in maternal labor—not just in paid work taken in (piecework or telework, as we will see in chapter 6) but in the very work of rearing children. Sleep is that site of reproduction that allows for the worker to get on, including social reproduction. Infant sleep poses its own problems, namely, that it is flexible but may be difficult to induce. But the history of infant sleep is also deeply interrelated to that of parents and other caregivers: the sleep of the child, and especially the infant, is in and of itself work, one that invites new ideologies and new tools.

II PREDICTIVE MOTHERING

3 HOT AND COOL MOTHERS

In 1948, *Time* reported on a new phenomenon: "frosted babies." Sensationally profiling Dr. Leo Kanner, *Time* wrote of the children that "they were apathetic, withdrawn, happiest when left alone. They shrank from anything that disturbed their isolation: noises, moving objects, people, often even food." Quickly, the article identified its target: the parents of these children—especially mothers. "The fathers were scientists, college professors. . . . All but five of the mothers had gone to college. . . . Cold Perfectionists."[1] The article interviewed the pediatric psychiatrist at some length, who remarked that the children were as if "kept neatly in a refrigerator which didn't defrost."[2]

Collapsing mother and child, container with contained, *Time* publicized Kanner's latest theory: mothers explicitly had something to do with the infantile autistic states he was diagnosing in his clinic at Johns Hopkins University. If mother was cold, so too would baby be. Class, education, and race (all of the families were white) each played their role in determining the conditions for diagnosis, but maternal intelligence, book learning, and regimens relating to a child were, he found, the most striking commonality and the foundation of his etiology. Issued into a scientific and social moment that corroborated this stance, the terminology leapt from medical discourse into the culture where it was wielded against such women.

Kanner was far from alone in identifying maternal traits and their impacts on child development at mid-century. Pediatricians, psychiatrists, and psychoanalysts aimed to identify what constituted a bad mother and how she produced bad children as a result.[3] This research was conducted in the United States and England by those from disparate schools of thought, some working in labs, others in clinics, some with theoretical children, others with actual children, and still others with live monkeys and other animals.[4] Several of these studies were conducted under the sign of behaviorism: ignoring the internal world of both mother and child to name the mother as a stimulus condition that leads to observable, diagnosable phenomena in children. This followed in part from John Watson's *Psychological Care of Infant and Child*, and flattened both mother and child, and the relationship between them, to an environment of inputs and their outcomes.[5] Other approaches were keenly focused on the fantasies, psychic life, or emotional home environments of children that produced pathological states, under the sign of either Freudian psychoanalysis or, later, attachment theory.[6] What emerged across the era's consideration of mother-infant relationships and child outcomes was a taxonomy that centered on the dosage and qualities of attachment, presence, absence, and affect—one that has been scientifically disproven, but which lingers culturally. Under this taxonomy, I argue, mothers provide either too much or too little stimulus for their babies, which in turn produces "undesirable" outcomes, or under- and over-affective states in children. These inquiries were frequently under the auspices of safeguarding children (especially sons), which meant that the primacy of the mother and her determining effects on children was a problem to be solved. On this understanding, a mother is always either too hot or too cool: the degree to which she and her affect are available was flagged as responsible for producing a divergent child. The baby is merely a passive receiver, a receptacle, and result—a symptom of a natural state of womanhood gone awry.

The most famous, if not persistent, of these "bad mother" theories-turned-diagnoses is the one with which I've opened—the "icebox baby," which, through a game of medical telephone, became known as the "refrigerator mother." Moving the emphasis on the chill from offspring to parent, this type of mother was understood to be one whose appliance-like affect is palpable in how she relates to her children and how her children then relate to the world. The refrigerator mother, its coiners claimed, produces autistic states in children by her ineffective and under-affective parenting. She was cool, and therefore generated a cool child, like begetting like. The refrigerator mother is not the only bad model mid-century psychiatry discovered; a whole host of other problem mothers, whom I call "hot mothers," was identified, as was a mother who ran both hot and cool by turns, like a tap. The lukewarm mother wasn't any good either. Almost any lived experience of the maternal yields a flawed outcome, a falling-off from an ideal, and any flaw can be traced back to dependency on a nonideal mother rather than to systemic forces. There is no right temperature, no right mother, only an ideal of natural, instinctive, healthy mothering implied by the theory but missing from its actual scene.

From the mid-1940s until the 1960s and beyond (because these theories, while discredited, have had powerful and violent discursive staying power in ableist, racist, and misogynist rhetoric), domestic technology and maternal function were linked by metaphors of temperature. Whereas autism and autistic states have been extensively elaborated in their relationship to digital media, I attend to the understandings of maternal etiology or the *cause* of "emotionally disturbed," queer, and neurodivergent children (as they shift from mother-child to media-child and back again). I argue that these newly codified diagnoses were inseparable from conceptions of race, class, and affect at mid-century, and were influenced by behaviorist accounts of stimulation, mediation, and domesticity. They reflect a set of theories of maternal absence and (over)presence whose echoes persist in our present in terms like "helicopter parent."

TAKING TEMPERATURE

Coalescing scientifically at mid-century, and entering mainstream knowledge in the United States by 1960 (during the Cold War no less), this hot and cool mother typology parallels both the rise of the suburban home and the image of the discontent, bored, and underutilized housewife at its center, surrounded by new appliances and entertainment media, as well as Marshall McLuhan's influential 1964 binary of hot and cool media.[7] McLuhan argued that some media are hot, providing intense quantities and qualities of information that engage multiple senses. He argued that

Figure 3.1
Marshall McLuhan with telephones. Photograph by Yousuf Karsh. Reproduced with permission of the Karsh Estate.

photographs, for instance, were hot, as in this photo of McLuhan himself (shown in figure 3.1).

By contrast, the telephone, he claimed, was a cool medium because it provides less information; embodied, in-person speech is a cool medium as well, though arguably hotter than a phone call. Film, hot. TV, cool. Paper, hot. Engraved stone, cool. As Nicole Starosielski writes, "For a medium, to be cold is to be *off*, to lack the ability to transfer information."[8] Heat media, by contrast, overstimulate their user. The upshot of this binary, which yokes media and information technology, is an implicit assessment of what each given medium *requires* of its audience experientially and affectively via engagement, but also what it *permits* the audience to do, think, and feel due to levels of stimulation. A cool medium, because it is comparatively underfeatured, requires *more* engagement and imagination to fill in the gap. A hot medium renders participation passive, absorptive. To quote McLuhan directly, "The effect of hot media treatment cannot include much empathy or participation at any time."⁹ For McLuhan, hot media produce cool subjects, and cool media produce imaginative, responsive subjects (hot). This is, at its base, a determinist media theory of stimulation.

The coincidence of the binaries is not worth exploring merely because they share a language and an era, nor because the affordances and effects of media are, I argue, also about under- and overstimulation in a receiver. If, as we've seen, mother and her surrogates are understood as a medium for care, Kanner's work literalized this formulation. Both McLuhan's binary of hot and cool media and the psychiatric binary of hot and cool mothering converge at temperature and are derived from culture changes and material shifts in domestic architecture and media (especially new home appliances and television) at mid-century. Where McLuhan argues that cool media are better because they stimulate less, neither hot nor cool parenting is any good. Beyond the related metaphorization of hot and cool, one's relationship to mid-century domestic media and technology culture was a way of thinking the etiology and pathology of non-neurotic diagnoses in children, namely, autism and schizophrenia. At its crux, then, the diagnosis of maternal (un) fitness and the sorting of good from bad media both rest on a theory of stimulation: the right kind, at the right level, accessed and deployed by the right people in the right homes. Yet where McLuhan seems to prefer cool media for the states they engender, there is no mother who provides the healthy amount of herself: she is either too much or too little stimulus.

Recourse to the metaphor of electronics, appliances, mechanics, and media in describing the problems of *mothering* and the resultant child resonates with McLuhan's theory of hot and cool media in ways that cannot go unremarked.[10] McLuhan is clear in this theory via the endless gestures he makes beyond his remit that anything can be classified as hot or cool based on what it imparts, not just media. People and cultures are also hot and cool,[11] and one can attribute this to the affordances of the object or person in question: how the receiver or audience or related term responds. As McLuhan extends his binary away from media to everything else, he produces a raced and classed binary of hot and cool—"backward countries" (cool) versus the "we" of the global north, which he terms "hot." While this may sound like the reverse of orientalist and colonialist narratives of the "untamed" or "uncivilized," it is not; the term "cool" in McLuhan's hands implies older, and "less advanced" society or, to use McLuhan's own word, "backward."[12] The racing and classing of the two problematic forms of mothering, hot and cool, often fall within this same binary, reversed: hot affects belong most typically to Black mothers, cool affects to white mothers. Hot affects belong to the poor and working class, cool affects to the middle and upper class. Both class and race became part of the criteria for diagnosis of a cool or hot mother, with massive ramifications: children who may have benefited from therapeutics only purchasable via diagnosis were often excluded if their parents—and therefore their race and class markers—did not fit the diagnostic criteria for a particular temperature. Put another way, parents determined the range of possible diagnoses for their children, much like, for McLuhan, a medium determines the response of its user. The diagnosis of hot and cool mothering implies a media theory of parenting.

LOST MOTHERS

It is no accident that the theorization of hyper and hypo states in mothers, and under- and over-mothering, emerged when World War II produced large groups of children separated from their mothers caused by the aftermath of bombing, mass death, and emigration. Especially in Europe, as well as in research conducted in the United States by refugees fleeing the Nazi regime, the states and institutions aiming to take care of displaced and motherless children could be used as laboratories for understanding the effects of maternal separation on children, distilling out the role of the mother altogether. Despite the absence of the father in nearly all of these scenarios, these psychological studies were almost totally concerned to determine the role of the mother, and her *presence or absence*, in a child's development. As we saw in the preceding chapter, war and its aftermath were hardly the only living laboratory for studying the mother-child relationship. Experimental study was also conducted in the prison as much as the orphanage, in the home as well as the clinic. The histories of childcare, child psychology, and child experiments sit uncomfortably close and, indeed, are intertwined, across this period.

Whereas René Spitz turned the hospital, the prison, and the orphanage into his labs (see the next chapter), and psychoanalytic study into scientific visual evidence, Harry Harlow, inspired by Spitz's work and that of Bowlby, attended to the very *necessity* of a mother in his experiments with nonhuman subjects at his primate laboratory at the University of Wisconsin–Madison. Almost accidentally arriving at the question of maternal separation and surrogacy in the breeding of a rhesus monkey population, Harlow's experiments were deeply in conversation with the work being conducted by Spitz and other psychoanalytic and attachment thinkers (like Anna Freud and Bowlby; see the next chapter), on the one hand, and behaviorists like B. F. Skinner, on the other (see chapter 2), who were more focused on questions of stimulus and behavioral outcomes. The experiments were controversial in terms of the extreme stresses that they put on live animal subjects: some were placed in solitary confinement, others were given a "mother" that dispensed nutrition, and some were left with their real mothers. Not surprisingly, those left in solitary confinement did much worse than those with even a fake doll as a "mother" figure. Of those with doll mothers, some were wire and some soft. The monkeys paired with the soft dolls did nearly as well as those paired with monkey mothers. Humanistic psychologists applied Harlow's findings—where the motherless monkey is more troubled psychologically than the one with the wire, cloth, or real mother—to their infant and child subjects.[13]

Less well-known than Harlow's were experiments that followed in their wake. In the mid-1960s, Robert B. Cairns and, later, Donald L. Johnson, psychologists at Indiana University, Bloomington, traveled to Springhill Stock Farms in Greenwood, about an hour north of the school. There they collected very young Dorset lambs, which were brought back to their laboratory. These lambs, separated from their mothers and then isolated from six weeks of age, were raised not by a wire mother but a *wired* one: the television.[14] Later, they were raised with collies.[15] The purpose of these studies was to deduce whether the lambs form an attachment to an appliance (albeit one that was both medium and message) and across species. They did indeed attach to television—becoming electric lambs—*and* to dogs. Both media and canine were seen, as Bowlby described the outcomes of the Cairns and

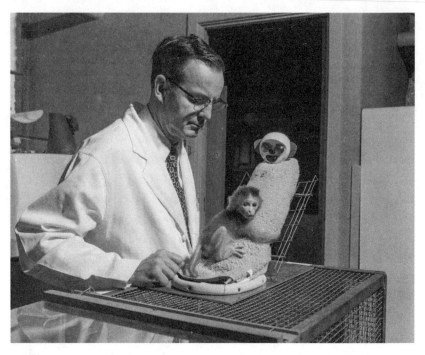

Figure 3.2
Harry Harlow conducting research on primates, ca. 1960–1969. University of Wisconsin–Madison Archives Photo Collection, S01464, Box 62, Folder 3/1, Harlow, Harry F., University of Wisconsin–Madison.

Johnson experiments, as "an attachment-figure." The lambs were in pain when either companion was withdrawn, "bleating on separation, searching for it."[16] Crucial here to the findings were that the lambs would attach to no more than an audio-visual stimulus, in the case of TV, or to dogs meant to herd, and thus hurt them. As with Harlow, in Cairns and Johnson's work the binary of hard and soft (and animate inanimate, loving or violent) proved less important than the binary of here and there, absent and present. Television made its presence known.

Sounding a great deal like D. W. Winnicott's formulation of the "good enough mother," this psychological research, like Spitz's, essentially found that any kind of mother was better than no mother at all. Put another way, we are dependent on mothers, but not only on those made of flesh and blood. Even a lifeless, wired, *cold* substitute is better than nothing.[17] The status of the mother, then, is reified in Harlow's work, but the primacy of the mother is her literal shape, her contours, her haptic stimulus, her embodied message. "Her" particulars are completely evacuated.

HOT MOTHERS

If the work of Spitz and Harlow (among others) proves we are dependent on mothers, and absent mothers are deeply damaging to the formation of a child, the question remains: What do specific kinds of mothers do and what is the impact of various forms of *presence*? Moving away from the warmth of a cloth mother and the cold of a wire mother, both better than no mother at all, mid-century psychologists and psychiatrists traced the effects of particular qualities of mothering through the closed circuit of their children. While the refrigerator mother may be the catchiest mother stereotype with the most staying power despite being out of vogue scientifically in our contemporary time, during World War II and after, figuring out the amount of maternal involvement that would neither overwhelm nor underwhelm a child with affect was seen as the key to avoiding pathological kids. By and large, while these investigators may have wanted to look for this ideal-in-practice, their looking was biased enough that they were never going to find it. What resulted was a group of diagnoses and stereotypes of mothers who offer "too much" (not explicitly sexually, although sometimes that, too). Psychiatrists claimed that mothers overwhelmed with demeaning tones, were too permissive, too attentive, too seductive, too masculine, too feminine, impacting their children by being unable to modulate their affects. As with McLuhan's hot media, they intimated that the excess of stimulation (whether in the form of infantilization, protection, or permissibility) begat different pathological states. Each specific subtype of hot mother produced her own subtype of pathological child.

Leo Kanner, who would later coin the term "autism," helped perpetuate this taxonomy in his elaboration of a new field: child psychiatry. Kanner had emigrated from Austria at the end of World War I and trained as a pediatrician and psychiatrist in South Dakota, where he wrote on idiosyncratic interests like the history of dental work and folklore, the antiquity of syphilis, and psychosis. In 1930, with grants from the Rockefeller Foundation and the Macy Foundation (in its very first year, and before it was associated with its famous conferences that in part turned their attention to the psychology and cybernetic theories of the child), Kanner made the first clinic for child psychiatry at Johns Hopkins. He rose to prominence by being vocally protective of mothers, railing against the societal pressures on them—especially those dicta about how to parent that were relentlessly issued by fellow pediatricians and psychiatrists. Whereas Spitz brought the empirical to traditional, Freudian psychoanalysis, Kanner wanted Freud killed off as the mid-century "Great God of the Unconscious." Kanner went so far as to write a book with a doubly pointed title, *In Defense of Mothers: How to Bring Up Children in Spite of the More Zealous Psychologists* (1941), in which he took a deeply anti-Freudian stance, and one ostensibly friendly to mothers—appealing to their innate common sense. He also derides mothers' "gullibility" and "obsessional" *labor* in striving to be good mothers via education, which Jordynn Jack contrasts with "libidinal mothering" or instinctive mothering: "Rather than rigorously quantifying their child's activities, the libidinal mother was to carefully observe her child's needs and to calibrate her responses accordingly."[18] Kanner takes on the quantified mother and demands she return to this qualified version. Yet his defense of mothers has less to do with valuing them—a feminist corrective to his era's male bias—and more to do with a rhetorical strategy for discrediting alternative theories of etiology.

Before Kanner would popularize the cold mother and her ice-boxy baby, he detailed the over-affective hot mother. In his *In Defense of Mothers*, Kanner's tone is not exactly, as one might hope or expect from the title, proto-feminist, nor is it uncomplicatedly on the side of mothers. Kanner and his more famous counterpart Dr. Spock uphold notions of common-sense mothering (as does D. W. Winnicott). But where, as we have previously seen, Dr. Spock would famously write just a few years later, as we saw in the previous chapter, in the opening to one of the best-selling books of all time: "Trust yourself." Kanner has the same hypothesis, with an execution at once similar and divergent. Although he begins his manifesto with an "Open Letter to Mothers," he quickly moves from directly addressing the mothers he seeks to defend, as Spock does, to a more distant third person. At once consoling and shaming, flippant as well as urgent, he treats parenting and its influences as a case study in psychopathology, much like the experts he derides treat children:

> There is no raid shelter from the verbal bombs that rain on contemporary parents. At every turn they run up against weird words and phrases which are apt to confuse and scare them to no end: Oedipus complex, inferiority complex, maternal rejection, sibling rivalry, conditioned reflex, schizoid personality, repression, regression, aggression, blah-blah blah-blah and more blah-blah.[19]

Nearly all of what Kanner references here is the popularization of Freudianism, the pediatric literature in an age of prescribed rigid feeding and toilet training schedules (as in the popular work of John B. Watson as well as the pervasive middle-class "scientific motherhood" of the early twentieth century), and the increasing popular market of advice books, into which *Defense of Mothers* would enter (and would go on to be taken up as gospel in a media panic over mothering, see figure 3.3). One of the greatest ironies of Kanner's

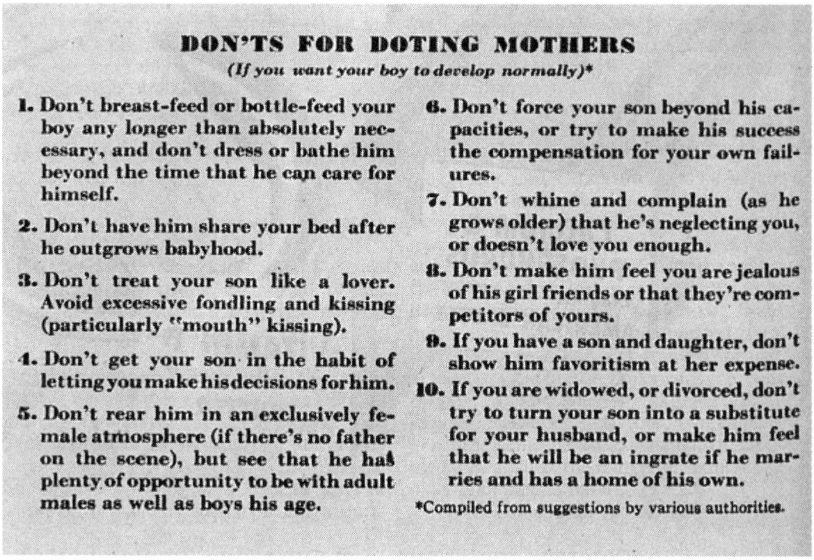

DON'TS FOR DOTING MOTHERS
(If you want your boy to develop normally)*

1. Don't breast-feed or bottle-feed your boy any longer than absolutely necessary, and don't dress or bathe him beyond the time that he can care for himself.
2. Don't have him share your bed after he outgrows babyhood.
3. Don't treat your son like a lover. Avoid excessive fondling and kissing (particularly "mouth" kissing).
4. Don't get your son in the habit of letting you make his decisions for him.
5. Don't rear him in an exclusively female atmosphere (if there's no father on the scene), but see that he has plenty of opportunity to be with adult males as well as boys his age.
6. Don't force your son beyond his capacities, or try to make his success the compensation for your own failures.
7. Don't whine and complain (as he grows older) that he's neglecting you, or doesn't love you enough.
8. Don't make him feel you are jealous of his girl friends or that they're competitors of yours.
9. If you have a son and daughter, don't show him favoritism at her expense.
10. If you are widowed, or divorced, don't try to turn your son into a substitute for your husband, or make him feel that he will be an ingrate if he marries and has a home of his own.

*Compiled from suggestions by various authorities.

Figure 3.3
"Don'ts for Doting Mothers," *Ladies' Home Journal*, November 1945.

life and work is that he ended up adding several terms to his long, pejorative list of complexes and psychological jargon: "smother mother," autism, and "refrigerator mother."

Kanner's description of mothering is also one of mediation—primarily the interference of print media, but probably psychological radio education as well, which was broadcasting heavily in this moment. Mediated mothering is the problem. Pure (good, white) mothering that comes from within instead of without is the solution. Of course, the very thing that impedes mothers (books—both their medium and their message) is supposed to correct the problem (Kanner's book). In thinking through the hot mother, the smother mother, he attributes, via class and therefore education level, the category error to *books*—which McLuhan will go on to call a hot medium. The hot medium of the book is a block on thinking for oneself; it misattunes the mother, and passes on its heat to her, converting her into a walking textbook that—in keeping with Kanner's notion that one type of mother begets one type of child—blocks a child's ability to develop into a subject that can think and do for itself. So goes the theory: media destroy maternal instinct and infect the ideally closed circuit of mother and child with overstimulation, which in turn causes the child to retreat from heat to coolness.

It is in this text that Kanner makes the link that problems with children stem from mothers—hot mothers, mothers who mother too much and oversubscribe to advice, mothers who are overwrought and hyper-anxious. There is not a trace of sympathy for why mothers might want to go beyond their own instincts toward the prescribed "good." He devotes an entire chapter to this type of parent, its sing-song name damning: the smother mother—a term whose rhyme contains twice the mother it should. Writing on the smother mother, Kanner does not mince words:

> Smother love is the most egoistically selfish thing on earth. It is a caricature of mother love. Its possessiveness is greed. Its aim is domination. Its logic is warped. It locks its treasure in a vault, away from circulation, and frantically expects it to yield dividends. The vault takes the shape of an imaginary uterus, which keeps encircling the child for years and years. It resists, prevents, prohibits, emancipation. It is sugar-coated cruelty. It treats the school child like

a baby, a college student like a kindergarten pupil. While mother love establishes ties, smother love forges chains. It commandeers a child's appetite, bowels, play, dictation, homework, everything. It strangely combines hugging with nagging, kissing with hissing, doting with don'ting.[20]

Kanner's smother mother is the definition of mothering to excess. As described above, her type is one who uniquely infantilizes her child, controlling them via mixed messages. Although Kanner does not elaborate on these paradoxical combinations, and does not describe the impact of this mixed messaging, hate is concealed in love (although one of his case children does remark that his mother couldn't have ruined him more if she had hated him instead of loving him). The anthropologist Gregory Bateson described exactly this problem as the "double bind": when someone, usually in family, is told to do two conflicting things, which are thus co-negated and impossible to negotiate. Drawing from theories of communication in the early 1950s, Bateson makes a typology of the double bind, and while it often appears innocuously in play, in fiction, in the structure of a joke, he also identifies a double-bind message as a fundamental structure in communications from parents to children (mostly from mothers, but, Bateson notes, not exclusively).[21] Kanner held his smother mother's excess responsible for a whole range of fragile mental states in children, while Bateson pinpoints the *duality* of her communication structures as pathological, writing:

> We have suggested that this is the sort of [double-bind] situation which occurs between the pre-schizophrenic and his mother. . . . When a person is caught in a double bind situation, he will respond defensively in a manner similar to the schizophrenic. An individual will take a metaphorical statement literally when he is in a situation where he must respond, where he is faced with contradictory messages, and when he is unable to comment on the contradictions. (257)

Bateson holds familial *communication*, via a double bind, responsible for emergent schizophrenic states, even causing those without a diagnosis of schizophrenia to react in a mode typical of that particular state.[22] Bad mothering is at its base not just a theory of media but a theory of communication as well.

Kanner and Bateson were hardly alone in diagnosing the problems of media and communications between mother and child. Just two years later, in 1943, the psychiatrist David M. Levy entered the fray. Levy had a storied career. By 1943, he was already famous for importing the Rorschach test to the United States in the early 1920s, coining the term "sibling rivalry," and developing new techniques of "active play" therapy in which children were encouraged to act out their aggression against dolls (including the mother doll—a literal remediation of mother). Levy had also served at the New York Institute of Child Guidance from 1927 to 1933 (which put out anti-media recommendations to parents; see the introduction to this book). Before Levy became famous in his diagnoses of mothers—more on this below—he wrote about the "primary affect hunger" in adopted children. Before World War II turned the human sciences' eye to the plight of mother-infant separation as a new territory for mother-infant interaction, Levy was interested in, as historian Ellen Herman writes, "how and why adoption might prove corrosive to emotional security . . . that adoptees suffered from its opposite."[23] Here he argued that mothers—and biological mothers at that—were essential. He then imported this work into his most famous text, his 1943 *Maternal Overprotection*. (During the Cold War era, adoption would only become more central to studies of mothering, especially transnational adoption.)

Alongside his paradoxical position on a board that rejected media for children and his real use of media in the consulting room, Levy wrote of both rejecting and overprotective mothers across this same binary in *Maternal Overprotection*. There, he argued, the hot, dominating mother produced an "infant-monster, or egocentric psychopath."[24] The book was successful and contributed to his bid the following year to start a new psychoanalytic unit, what would become the Columbia University Psychoanalytic Clinic for training and research (later, he would become the president of the American Psychoanalytic Association).[25]

Just two years after Levy's writings on the subjects, and five after Kanner's thesis debuted, Edward Strecker advanced this over/under maternal stimulus theory further. Published in 1946, *Their Mother's Sons: The Psychiatrist Examines an American Problem* is an extension of a lecture Strecker had given just the year before to hundreds of medical students and doctors at Bellevue Hospital, entitled "Psychiatry Speaks to Democracy"; the notorious lecture would go on to be called simply "The Mom Lecture," eliding its macro concern. In that lecture, as well as in the articles that reported on it ("Are American Moms a Menace?" in *Ladies' Home Journal* and in the *New York Times*, Strecker advanced the argument that the apron-strings form of mothering (overly coddled, overly attached) was ruining democracy: the maternal sin against the child was a sin against America.[26] Substituting the primacy of the mother for that of the motherland, Strecker used newly available hard data to back up his claim: the mass surveillance conducted on American men in the form of the draft board—with 49 million men registered.[27] Adolf Meyer, in a coup for Strecker, wrote the introduction. Meyer, who had a hand in training a great number of the "mother-as-medium" theorists (see the introduction to this book), and had given Kanner his Johns Hopkins job, was then the chairman of the National Committee on Mental Hygiene and a founder of the mental hygiene movement in the United States. Meyer wrote in his preface to the volume, "To him the cold hard facts that 1,825,000 men were rejected for military service because of psychiatric disorders, that almost another 600,000 had been discharged from the Army alone for neuropsychiatric reasons or their equivalent, and that fully 500,000 more attempted to evade the draft were alarming statistics."[28] From there, Strecker hypothesized that this huge number of men unfit for duty, from all walks of life, shared one problem: bad momming. "Momism" had already appeared in Philip Wylie's 1942 screed, *A Generation of Vipers*, wherein he differentiates a mother (good) from a mom (bad). Although not a psychologist himself, Wylie's book would become synonymous with the "bad mother" panic and was in its twentieth printing not even fifteen years after its release. (As an aside, Wylie's own daughter ended up married to the second Lindbergh baby and was a behavioral psychologist herself, writing about human chestfeeding and marine mammals, and developing clicker training—the positive reinforcement technique used with household pets.)

Strecker took this nascent idea of "momism" and brought it back to psychology, taking it further and generating an entire classificatory system for the pathology of moms and momism.[29] As in Kanner's book, the deployment of war metaphorics, as well as scenes, is extensive. Strecker demonstrates the high stakes of bad mothering with endless anecdotes from war rehabilitation centers, most unfolding in the pattern of: if I hadn't been such a coward, my "brother" (another soldier) would be alive.

As many scholars have argued, bad mother theories were deployed in the service of safeguarding democracy and shoring up future democratic subjects via the psychical health of their mothers; problems at the micro level of the family were scaled up to the level of the nation, both in the United States and in England. As Deborah Weinstein writes, "Because the family was seen as a key site for the production of healthy personalities and productive citizens, child-rearing patterns held the potential to bolster or undermine American politics and society by affecting the psychological well-being of children.... [B]ad parenting became the cause of fascism, prejudice, autism, and homosexuality."[30] Strecker's title and the opening of his lecture nod at this contemporary work on the macro problem parenting produces (a failed, emasculated generation of men, as well as the next in the process of growing up), before taking a turn to unvarnished diagnostic misogyny focused on the failure of women. The book that follows is a tour de force of mom typing (there are seven kinds) and their terrible outcomes. Each has a form of a "silver cord" that replaces the umbilical cord as soon as it is cut, an extension of mom.[31] This cord is another loud metaphor—presaging Norbert Wiener's *Cybernetics*—that replaces the organic with the inorganic, intimate flesh with a metallic, machinic, and cold form, much as Harlow had done with his monkeys and their wire mothers. Strecker turns the umbilical cord—a site of perfect, attuned nourishment—into a power source and a chain. It takes the site of the cut cord—one that allows for a first differentiation between mother and child—and reinstates a bad connection, pathologizing attachment that, under this logic, becomes unbreakable.

This image of the metallic, unbreakable bond is a metonymy for bad moms: moms who are so controlling that they are to blame for literal fascism, moms who are so doting they are to blame for inept children, for submissive children. Strecker's text is a critique of maternal femininity—vain femininity, masculine femininity—and its effects on masculinity (as per the title; Strecker did release a sequel called *Her Mother's Daughter* ten years later).[32] An entire chapter is devoted to what for Strecker may be the worst outcome in a son: homosexuality. As Jennifer Terry writes, Strecker "cautioned mothers to loosen the stifling grip of maternal love and allow sons to overcome the Oedipal crisis in order to become healthy, masculine heterosexuals."[33] Strecker's homophobic theory, which frequently confuses gender for sexuality, admits that there may be a neurological or biological "deviation" responsible for *some* gay men, but that others are only gay because their mothers wanted daughters and feminized them, or smothered them (as Kanner describes).[34] The silver cord and its maternal power supply are always to blame (and he is far from alone in this theoretical assertion—many psychoanalysts and psychiatrists blamed mothers for gay children on the grounds of a monstrous intimacy and overidentification; in the 1960s these logics reappeared on the grounds of the trans child and their excessive mother).[35]

Under Strecker's conception, the problem with mothering was exacerbated by a top-level societal change: a shift toward, in his own words, *matriarchy*, and a general absence of fathering. This was, in parallel, articulated along the lines of postwar women's advancement; even in the late 1940s, as Ellen Herman shows, feminism became understood as a form of maternal illness.[36] Compliant mothering—not women's emancipation—was the aim; the recommended cure was widespread psychotherapy. We can add to this account that women's sexuality—especially what it meant for children that mothers might be sexual—was in great flux in this moment. While "hotness" was long in use to mean sexual, it became slang for *being* attractive rather than "hot" *for* someone (which remains its other slang usage) in the interwar period. After the best-selling, landmark publication of Kinsey's second volume, *Sexual Behavior in the Human Female*, in 1953, Cold War ideologies about a warm home, and a warm wife, mixed with nascent notions of a new sexuality, begetting new forms of attractiveness and sexuality in terms that were thermonuclear: the "bombshell," or being sexually powerful like a "nuclear bomb."[37] But suspicion, as Lynn Spigel and Elaine Tyler May show, was attached to single women—as well as single men in the Cold War. A woman needed to be sexually active and enthusiastic, but only for her husband.[38] This too became a double bind.

Under contemporary logics of racialization, the theory that hot mothers, whether bored and thus overfocused on their child, working outside the home and thus masculine, or offering a perverted feminine resulting in smothering, delinquency, ineptitude, and even queerness persisted, attached to Black mothers. This happened in parallel to pathological cool mothering affects (long belonging to those white, educated feminists, from the Freudians' frigid women to Kanner's refrigerator mothers) becoming increasingly attached to whiteness in the 1950s and 1960s. This was by no means the first raced and racist "bad mother" theory attached to Blackness: as Stephanie E. Jones-Rogers shows in *They Were Her Property*, this theory has its socio-medical antecedents in the paradoxical dependence on (and pathologization of) enslaved women laboring as both wet nurse, surrogate mother, and possible site of cross-racial contamination for white children in the antebellum South (see chapter 1).[39] In the context of the twentieth century, the interwar period was marked by progressive and liberal studies on prejudice and its effects on Black families, as Ruth Feldstein and Daryl Michael Scott show in their works. It was, instead, in the postwar era that psychiatrists and social scientists pathologized the Black family as disorganized and generative of pathological children. This was intimately tied up with the critiques of "momism" advanced after the war, and only concretely emerged when "medicalized notions of personality" were taken up readily across the social sciences.[40] With this medicalization came the study of "damaged personalities arising from black family life . . . virtually everyone agreed that 'matriarchy' had adverse consequences for the personalities of black people."[41] The Black mother was most frequently understood to be a bad mother, in part because she was alone and the sole parent: she was intrinsically dominant, a hot mother, even though rarely present (she was always described as away at work), emasculating her children.[42]

This is perhaps nowhere more boldly, schematically articulated than in the 1965 Moynihan Report, which, via its recourse to psychoanalytic theory, goes so far as to say that Black parenting is reducible to Black mothering—there is no such thing, in reality, as a Black father.[43] That absence, according to Moynihan, has many ramifications, including that mothers therefore work, and deprive their children of attention (under-stimulating them in terms of education), while being overall too domineering and disciplinary, resulting in delinquent children—again, mostly sons.[44]

Under the guise of a liberal report to justify new racial equality policies, the report furthers racist bad mother theory. As Kathleen Stockton writes on the Moynihan Report, "So strongly does the report believe its claims . . . that they become the frozen, mythological rendering of black families for all time."[45] Hortense J. Spillers's critique of the report is multifaceted, and includes its erasure of Black fatherhood, which ensures that, yet again, the mother is held responsible for the misfortunes, misattunements, and problems of her sons.[46] Unlike white families, whose mothers are to blame *as mothers* in spite of their sons belonging just as much to their fathers, Spillers argues that "the 'Negro Family' has no Father to speak of—his Name, his Law, his Symbolic function mark the impressive missing agencies in the essential life of the black community, the 'Report' maintains, and it is, surprisingly, the fault of the Daughter, or the female line."[47] The consequences, as Kevin J. Mumford shows, were to elaborate a notion of the stable white family, a Black family in disrepair, and center outcome of homosexuality in children.[48] The absent white father, away at war, of course goes unremarked because, as Spillers argues, white sons belong to their white fathers even in absentia, although their mothers are responsible for their ruin.[49] Via the kinship structure mythologized in the Moynihan Report, Black mothers are doubly to blame for the lack of Black fathers, and for all possible child outcomes via their heat.

COLD MOTHERS

Whereas Spitz, Harlow, Freud, Bowlby, and others at mid-century focused on theories and resulting mental states of children physically deprived of their parents, and Strecker criminalized the "American Mother" as the site of reproducing pathology, Kanner moved from railing against psychological advice and its influence on parents and over-presence to the impacts of another kind of maternal flaw: emotional withdrawal. In 1943, right on the heels of *In Defense of Mothers*, Kanner famously elaborated a diagnosis of "early infantile autism," also called, as these things tend to go, Kanner's syndrome. Kanner's new diagnosis of infantile autism served as a double revision of widely held understandings in pediatric psychiatry, as well as his own publicly elaborated thought on childhood mental states. First, prior to Kanner's elaboration of infantile autism, the state he observed and described had been correlated with schizophrenia; the social withdrawal he classified as part of autism had previously been thought of as a symptom of psychosis.[50] It would take more than twenty more years for this shift to be reinforced completely in psychiatry.

In his 1946 study, Kanner observed three girls and eight boys, all of whom were parented by those in the local community. These were his "ice box babies." It bears repeating that, rather than yielding a diverse, if miniature, sample size of those in Baltimore, these children were white, belonged to the middle and upper classes, and were the children of "highly intelligent" parents, including several children from the professoriate at Johns Hopkins, and at least two children who were offspring of Kanner's medical colleagues.[51]

What Kanner observed in the children was a turning inward and sometimes a replication of sociality but not sociality itself. As Amit Pinchevski and John Durham Peters note, one of the criteria that Kanner began to develop was a rejection of the social and a gravitational pull toward *things*, namely, objects of play: "Objects supplied what people could not: predictability and monotony. Kanner portrayed his patients as having a uniform *modus vivendi*, relating to everything and everyone equivalently—as objects."[52] When the children preferred objects, Kanner began to describe their mothers as objects too.

Kanner theorized that there were, in addition to likely natural causes, environmental causes for this state, which he called autism. Several things followed. The first was that Kanner observed that the *mothers* in these families were never deeply emotional. Instead, they were reserved, removed, and with low affect. In describing how the mothers tended to the children, Kanner writes that it was "the mechanized service of the kind which is rendered by an over-conscientious gasoline station attendant"—again a cord metaphor for providing just enough fuel to go—but no more.[53] The fathers were deemed too busy to take notice of the children—which also meant that they were home and not abroad fighting in World War II (and thus free from the draft—another marker of education and class level or, following Strecker, a defect of character and a deficit of masculinity). The second thing that followed was that, in his observation of the narrative history of the children, he linked all eleven children's class background, and their origin in white families, directly *to* the children's autistic states. The result: children who appeared as if they were "kept neatly in a refrigerator which didn't defrost."

The cold, white, middle- or upper-class mother was yet again to blame. It followed that there was no proof autism was possible in children from "unsophisticated" parents, by which Kanner of course meant lower-class parents, parents of color, parents without a secondary education, or an admixture of all three. As Jordynn Jack observes in *Autism and Gender*, the race-based medicine Kanner practiced was not only proper to his particular research on autism but became the prototype for causation.[54] Medical racism, which limited access to intervention and study in 1940s Baltimore, laid down a fundamental inequality in attention to autistic children of color, the legacy of which continues today. Additionally, Kanner's revision of the long-standing belief that autistic states were a symptom of schizophrenia was part of a larger trend toward differentiating adult schizophrenia (with an adolescent onset) from autism (with an early childhood one); this had the further consequence of moving schizophrenia away from its historic diagnostic association with white femininity to its ongoing association with Blackness and Black patients starting in 1960.[55]

Of schizophrenia, Jonathan Metzl writes that in the first half of the twentieth century, the typical patient was seen as an "unhappily married, middle-class white woman whose schizophrenic mood swings were suggestive of 'Dr. Jekyll and Mrs. Hyde.'"[56] These conceptions reversed, and schizophrenia became attached not to "docility but to rage" and especially to Black rage in the form of Black emancipatory activism.[57] But there was a concurrent force at work that needs to be similarly accounted for. Kanner was not by any means the only psychiatrist in the United States to work with autistic patient populations in the 1940s. Lauretta Bender, who had, in part, trained at Johns Hopkins and was a former colleague of Kanner, had by this moment already described autistic states (but not by that name). In the 1930s, Bender worked with her child and adolescent patients at Bellevue Hospital in New York, and thus treated a largely Black population. Bender had already presented her studies publicly—work Kanner was intimately aware of and took as the inspiration for his investigations at Johns Hopkins. But Kanner took the limit of his population as the de facto diagnostic criterion for his syndrome.[58] To put it bluntly, cool parenting affects were the domain of whiteness (autism/distance/under-stimulation); hot parenting affects became increasingly associated with Blackness (schizo/split/hyper). Autism became a white illness while schizophrenia became a Black one, and I contend that these shifts are linked as the diagnoses become distinguished from one another. This medical redlining was so complete that Black mothers, such as Dorothy Groomer, were *unable* to get their children diagnosed as autistic, specifically because they were Black (and therefore were excluded from treatments for autism). When Groomer sought a diagnosis for her son Steven (via Bruno Bettelheim; see below), she was told it was *impossible* that her son was autistic; instead, he was "emotionally disturbed."[59] Groomer understands this as being because as a family they didn't fit diagnostic criteria, even if Steven did. As Groomer puts it, "This was not a negotiable issue. . . . They said, 'You can't even be a refrigerator mother.' The irony of it all."[60]

Kanner's notion of a frosted (white) child—the icebox baby—migrated etiologically to the "refrigerator mother." In his treatment of mothers and resulting children, Kanner metaphorically linked the child's pathological removal to an appliance-like mother—a *new* suburban appliance—and to fueling a car, turning the child into a machine, and the mother into oil (a medium). Taking Kanner's work, both before his elaboration of autism and after, together, mother-infant interaction is too hot or too cold—she is over and under affectively present. Mothers are seen as a function, not persons, and thus are easily metonymizable to domestic architecture and other helping functions. Because the maternal function is read as stimulus and not as care, it is also continuous with media and their inputs and outputs. The use of hot and cool is not a coincidence or a neutral media theory of parenting, but a misogynistic and racist media theory of parenting that depends on reducing the mother to a mechanized stimulus transmission. Further, the withdrawal, coldness, unnaturalness of the mother on this understanding, her absence while present, may be tied to women increasingly working during the war (while Kanner is conducting his study) and after it, when the "refrigerator mother" theory was codified. Women who were both wives and mothers had their work outside the home criminalized by the same psychiatrists who were pathologizing their mothering styles while they were in the home. In the postwar era, a mother who worked outside the home was blamed for the rise in delinquent children and other negative types. Her work inside the home was recast as unnecessary due to the new appliances that "helped" her with it—she had no reason to be exhausted, withdrawn, or emotionally absent.[61] What remained for her was to focus exclusively on her children, which was its own problem.

Shifting conceptions of class and women's labor haunt the new diagnosis of refrigerator mother. Ruth Cowan writes in *More Work for Mother* that industrial modes of domesticity evolved over a full hundred-year period, from 1860 to 1960. For some, this process was gradual and for others it was a rupture. Cowan marks the postwar period of 1945–1960 as its own distinct moment in which an endless proliferation of appliances inflected notions of class, affluence, consumption, and labor for those performing domestic housework.[62] By 1960, nearly every American (there were exceptions) lived in an industrialized domestic space. As Cowan puts it, "the diffusion of affluence meant the diffusion of toilets, refrigerators, and washing machines, not Cadillacs, stereos, and vacation homes."[63] By contrast, in 1940, she notes, a third of Americans drew water outside and carried it in in buckets, and two thirds were without central heating. Even by 1945 this had changed. Cowan famously argues that managing the increasingly electrified domicile was not, as is popularly held, easier for women charged with maintaining the home. It was, as she claims, all a sleight of hand; time-saving was time-increasing, and mother bore the brunt of this shift. The management of children and childcare at mid-century changed in response to all the duties pinned on the worker we call the housewife. It was this development—the sudden ability to own domestic technologies and to augment mothering via technology and para-mothering—that the refrigerator mother theory targeted while attaching race and class to specific disturbances in the mother-child relationship at mid-century. One more particularly relevant example: in 1941, only half of Americans had access to a mechanical refrigerator; by 1951, that number was 80 percent.[64] The refrigerator mother theory, however unconsciously, however metaphorically, records a shift in the role of the mother at home, in her new environment, with her latest devices. Domestic media entered into this equation as well (see chapters 5 and 6).

More than Kanner himself, Bruno Bettelheim popularized the refrigerator mother theory, yoking it even more firmly to domesticity at mid-century until it became the dominant, derogatory term for the mother of an autistic child (it was also open, at least some of the time, to reappropriation as evidenced by parent attendees at the first meeting of the National Society for Autistic Children, who wore nametags shaped like refrigerators, turning it to a literal self-appellation).[65] Bettelheim used his media appearances, which were various—from his cameo in a Woody Allen film (where he speaks briefly on the nature of conformity in *Zelig*) to stints on mainstream television, including multiple visits to the Dick Cavett show, to broadcast this mid-century pairing of autism and the refrigerator mother and was so vocal about it that he is sometimes incorrectly referred to as its originator. Bettelheim made sure that the refrigerator mother was perhaps the most famous of psychological concepts in that laundry list Kanner wrote up in *In Defense of Mothers*—on par with the likes of Freud's Oedipus complex and Jung's Electra complex.

Bettelheim, also an émigré from Vienna, arrived in Chicago in 1939 after having been interned in Dachau and Buchenwald. Bettelheim is a multiply scandalous figure in the history of psychiatry. He perhaps never earned a clinical degree in psychology or psychiatry (and instead had one in the history of art—at a time when the United States only allowed medical doctors to immigrate and refused entry to lay analysts fleeing Europe). His first work was to describe individual and group psychology in extremis, much like what he had encountered in the camps. That article, published in 1943, turned out to be controversial, and much of its evidence falsified.[66] Nonetheless, in part on the basis of that article, Bettelheim became the head of the Sonia Shankman Orthogenic School in Chicago, where he worked with many kinds of "emotionally disturbed" children,[67] including children with the newly emerging diagnosis of autism. As Chloe Silverman writes, the Orthogenic School was unique for its time—both architecturally beautiful and less restrictive: "The children could leave the grounds at any time, but visitors needed permission to enter. The institution protected those living there from real and imagined threats outside the gates, even as the staff encouraged independence."[68]

In his work at the school, Bettelheim built on Kanner's cool mother to elaborate more fully the refrigerator mother's suburban existence as cold and bored (i.e., under-stimulated), reflective of her marriage and the domestic setting in which she found herself.[69] Notably, this typology and the connection it forged between the effects of industrialization and its domestic technologies on mothers and their states of mind (including boredom) has its roots in the interwar period, where it was first attached to *hot*, domineering mothers at loose ends, and would continue to be engaged extensively in the culture via Betty Friedan's 1963 book, *The Feminine Mystique*, for a very different purpose.[70]

In addition to popularizing the appliance metaphor, Bettelheim's scientific work reached a wide audience, allowing him to publicize the work he did at the Orthogenic School defining autistic/schizophrenic states (again, the diagnoses were coterminous for some and not for others until the mid-1970s when they were fully distinguished; for Bettelheim, this happened in the mid-1960s). Of that research, perhaps the most well-known is his case study of "Joey," published in *Scientific American* in 1959. Joey is himself the refrigerator, so to speak, a "mechanical boy," an appliance. The case study has been much written about precisely because of Joey's self-identification as electric, which puts Joey, in some ways, more firmly in the lineage of schizophrenic patients who have a special connection to feelings of electricity and being electrified, from Daniel Paul Schreber (whose book *Memoirs of My Nervous Illness* became famous when Freud turned Schreber into a case study for his exploration of the paranoid schizophrenic) onward.[71] Joey, nine years old and autistic, had a specific relationship to the electric: like an appliance, he had to be "plugged in" to "work" and to survive. His "fantasy" or "delusion" was so complete, Bettelheim reports, and so persuasive that one could almost believe that he was partly if not wholly robotic. Bettelheim relates that even the fellow children and the maids of the Orthogenic School took special care and attention not only with Joey, but the playthings involved in his fantasy of being a machine, understanding that he needed his "wires" to live and his "carburetor" to breathe.

The case study can, of course, be a violence, where the descriptors and scenes are narrowed and contorted to fit a working theory of its author rather than reflecting the lived experience of its subject.[72] Given Bettelheim's record, this may well be so in the case of Joey; in one sentence Bettelheim owns this, writing, "His story has general relevance to understanding emotional development in the machine age."[73] This sentence is stranger than it appears, precisely because Bettelheim goes on to pin Joey's withdrawal on his mother in the familiar ways. His mother was in deep denial about her pregnancy. Her relationship to Joey did not change upon birth. Whereas Kanner observed children whose fathers were somehow untouched by *going* to war, Joey's father is a veteran described as "rootless" and unprepared to parent.[74] Instead of distributing causality across both parents and the structure of the family, or on families reuniting after the war, Bettelheim derides the lack of attachment in Joey's mother while also making her exemplary: a type and a pathology.

Bettelheim's focus on Joey's mother is layered. First, Joey was maintained on a rigorous schedule for feeding and toilet training so that he would be as unobtrusive as possible. While Bettelheim clearly states that this is common—children are trained and monitored—his parents were extreme about Joey's scheduling. In essence, his parents turned him into a machine, a machine that ate, drank, urinated, and slept on a fixed, automated timetable. Of this Bettelheim writes, "His obedience gave them no satisfaction and won him no affection or approval. As a toilet-trained child he saved his mother labor, just as household machines saved her labor. As a machine he was not loved for his performance, nor could he love himself."[75] Throughout the text, Bettelheim attributes the lack of emotionality not just to Joey's mother but to the "comfort so readily available" in the new domestic setting of the 1950s.[76] Bettelheim here folds domestic architecture into a critique of the mother, when the child is the one acting like an appliance (even if his mother presents as a classic example of the appliance-based mother "type"). Given the ease of the life he assumes for Joey's mother, Bettelheim diagnoses her as totally pathological for not finding her way out of her malignant ambivalence toward Joey: she has nothing else to do but love correctly.

When Joey first presents to Bettelheim, he only says "Bam" when addressed—a mechanical explosion. Bettelheim writes, "Joey plainly wished to close off every form of contact not mediated by machinery."[77] Not just functioning machinery (though he made those sounds and gestures too, churning like an engine), but *broken* or exploding machinery. Bettelheim interprets the fantasy of being mechanical as a wish to be stronger, better than a human, precisely because being a human and needing was "too painful." Bettelheim argues that Joey turned to machines because humans—namely, his mother—did not nourish him on any score. So, Bettelheim argues, "Joey's delusional system was the artificial, mechanical womb he had created and into which he had locked himself."[78] Bettelheim does not get there, but it may be that all of Joey's wires are umbilical cords, retethering him to a mechanical mother or to a more reliable substitute, much like Strecker's punitive silver cord. For Joey, becoming electric allows him to become self-sufficient yet perform a reassuring dependence on a more reliable power, even if from a cold, nonhuman source. Electrification requires both attachment/connection and output/reception/output. Joey makes a closed circuit within his own fantasy, even if the fantasy relies on this outside electrical source.

The Joey case, despite its various forms of extremity, from the biographical detail to the way Bettelheim unfolds and interprets it, contains so much of the motives and assumptions of psychiatric diagnosis at mid-century. Bettelheim reduces mothering to function—that of providing stimulus—and, taken together with the more extensive literature on bad mothers, describes maternality as crucially causative but always going wrong. At the same time, Bettelheim both overtly and covertly argues that a mother—especially one aided by access to the abundance of mechanical or para-parenting assistance in the home—should be not just a function but a primary person. This is a structural double bind. The mother is neither present enough nor specific enough in her stimulus, in her attention, to the child. In effect, the scientific description is self-fulfilling, accusing mothers of its own metaphorics. Put another way, the mother is the metaphor.

Kanner, Strecker, and Bettelheim comprise by no means an exhaustive inventory of theorizers of the *cold* mother, nor was the phenomenon contained to mid-century, nor only as a set of pathologizing prescriptions. Hot and cool mothers show up across the remainder of the twentieth century into our present. Feminist psychoanalysts, by way of Melanie Klein, Karen Horney, and Maria Torok (among many others), have investigated both individual mother-child dyads as well as the social reproduction of mothering since Freud. Nancy Chodorow's hallmark psycho-sociological works on this topic in the 1970s and 1980s expressly deal with the relationships formed in reaction to dominant (hot) maternality, and its effects on children, especially daughters. André Green, the celebrated French psychoanalyst, elaborated in the 1990s the "dead mother complex"—again as a tool to explain a resultant state in children because of particular maternal depressive affects. The dead mother complex describes what happens to a young child when a mother, initially lively and loving, becomes unreachable through extreme depression. Green describes this mother via now familiar metaphors of temperature and mechanics. She is suddenly "switched off" and has a "cold core" much like a lake of ice. Similar to the refrigerator mother theory, in which a cold mother produces a cold child, Green's is also a theory of replication, but via a psychical process—identification—wherein the child mirrors the mother. Nevertheless, the mapping of maternal *parenting* to cool devices, environments, and states (the icebox, the refrigerator, the banal sociopathy of a Nazi, the lake of ice) continued to be propagated.

FROM COOL MOTHER TO HOT MEDIA AND BACK

In our contemporary moment, both social and medical diagnoses have begun to substitute out the mother figure for media as site of etiological blame for autistic states, pathological antipathy, and other diagnoses of divergence and disturbance. Whenever discourses about media are added into the mix, the hot and cold mother and her hot and cold children are necessarily present. Despite the gains of disability activists and mental health professionals working to undo the stigma and ableism that run rampant culturally as well as scientifically, neurodivergent children are still to be "corrected" via these working theories (for instance, that screen time is to blame for an increase in autism diagnoses as opposed to, say, early intervention screening and shifting diagnostic categories).[79] Scholars have attended to this blending of media theory and usage and autistic states from a wide variety of perspectives.[80] It is not as if we have ceased to figure the mother altogether as site of relation that produces these mental states and substituted for her with the screen, because the mother is still tethered to the uses of the screen, and endless debates about screen time and permissive parenting permeate American culture in the twenty-first century. This latest joining of mediation and maternality can be read through the lens of the work at mid-century of diagnosing stimulation and states too hot to handle or too cold to love.

The helicopter mother has many snide variants: snowplow, bulldozer, lawnmower, submarine—once again, the coinages draw on the nonhuman for their classificatory metaphor, most frequently machines, to describe and pathologize qualities of parenting and to account for generational "problems" and qualities. As opposed to mid-century metaphorics, which centered on temperature and domestic, interior appliances, all of these contemporary mother types are figured as outdoor machines, but from the point of view of maternal typology, it is more important to note that they are non-domestic machines. Like "wearing the pants," the difference in appliance registers the shift from a suburban housewife at home with her refrigerator to the defeminizing role of full-time work outside the house that takes place first in the 1940s and then again in earnest from the 1970s onward. These stereotypes, much like smother mother and refrigerator mother, are deeply attached to ideals and fantasies that are classed and raced.

The helicopter mother first makes "her" typographical appearance in an off-handed remark by a teenager (a son) who describes his mother via this metaphor in the 1969 parenting book *Between Parent and Child*.[81] Despite this early usage, helicopter parents are understood to be distinctly opposed to the hands-off working mothers of the 1970s and 1980s, or even their opposed complements in the same period, attachment parenting advocates (see the previous chapter).[82] The helicopter parent did not emerge as a distinctive type—and the site of intense social and psychological scrutiny—until the mid-1980s and early 1990s, when helicopter mothers were differentiated from attachment parenting, and formally coined and tied to generationality: boomer parents hovering over their millennial children.[83] The casual, shaming diagnosis continues to be debated in our present, as millennials themselves become parents. Though the term was at first restricted to the mother, starting in 2000, helicopter *parents*—a non-domestic machine but a form of surveillance—entered general usage. In the present, conceptions of parenting have moved further away from mothering and toward a more gender-neutral representation, not only because of forms of social progress but because of the fate of describing parenting (mothering) as function; it can become gender-neutral or plural because it is once more described as a nonhuman stimulus source. This quality of parent is defined by *presence*— but again, presence of what kind? A ubiquitous presence but at a distance, overstimulating but *under*attached.

Ironically, helicopter parents are often spoken of as if they are *too* attached and unable to let their children go differentiate—the opposite of the mothers Spitz worked with in the mid-century prison (as we shall see in the next chapter). Using attachment here, in the psychological sense, is a misnomer, both in its recollection of attachment parenting as its own distinctive philosophy and in its association with attachment theory (in the works of Bowlby and Ainsworth, who separately and together elaborated experiments and understandings of what makes a child secure—secure enough to attach without avoidance or anxiety and thus develop and present normatively). While attachment parenting and helicopter parenting may share the hallmarks of high emotional support, helicopter parents are often spoken about as purposefully controlling their children's behaviors and arresting the development of autonomy. They are not a base from which to explore the world; instead, they are figured as chasing after their children, clearing the path in front of them, operating behind the scenes, and omnipresent.

The helicopter parent belongs to the domain of whiteness, and is almost always upper-class and highly educated, much as was the case with the refrigerator mother. Nearly identical in desired outcomes for her children is another "hot mother" type—the tiger mother—an Orientalized (as well as self-claimed) and machine-free appellation for a usually Asian or Asian American mother.[84] Both these types are the twenty-first-century reprise of the smother mother, similarly demarcated into category by race and class. As Malcolm Harris points out in *Kids These Days*, this classification erases the maternal labor that working-class and middle-class mothers perform as advocates for their children;[85] many critics who chide helicopter mothers focus exclusively on the parenting behaviors and outcomes of college-educated middle-class and upper-class mothers, while altogether ignoring other kinds and qualities of intensive parenting performed by mothers in other race and class positions.[86]

The backlash surrounding these mother types reprises nearly exactly the "momism" discourse surrounding hot and cool mothers at mid-century. The psychologist Madeline Levine, who writes about these parenting types, accepts this taxonomy and their traumatic impact on their children and teenagers, and argues that for all the hovering, monitoring, surveilling, and pushing, the parents are not *emotionally* present to their children. Much like the over-educated smother mother who relied on parenting texts rather than her own maternal instinct, helicopter parents are derided because they are over-present and use external benchmarks as a parenting guide; these are the middle- and upper-class parents of late-stage capitalism, who are stereotyped as believing that a child's acceptance into an Ivy League school is the be-all and end-all, the crowning parenting achievement. As Hara Estroff Marano argues in her 1994 book *A Nation of Wimps: The High Cost of Invasive Parenting*, "hothouse" parenting, used interchangeably with helicopter parenting, is to blame for widespread social ills.[87] Despite the fact that the jacket copy trumpets that this is the "first book to connect the dots between overparenting and the social crisis of the young," Marano's stance and rhetoric are nearly interchangeable with that of Strecker and Wylie some fifty years earlier.

If the helicopter is one technological metaphor for this form of parenting, it has also been attached to notions of stimulation and communication: "always on," like a smart phone. Instead of practicing attachment parenting, in which parents consistently *physically* hold young children, helicopter parents hover, making use of adjunctive tethers like texting, which Richard Mullendore has described as "the world's longest umbilical cord." Reminiscent of the folksonomic apron strings and Strecker's silver cord, the smart phone and surveillance apps are the latest problematic yoke of contemporary parents to their children.[88] The criticism argues that persistent contact is not the same as attention. If these are examples of hot parenting, their heat is cold.

Helicopter parents are understood to surveil and interfere, and that surveillance can be read as either hot (too much control) or cold (too distant) or both.[89] Surveillance implies distance and, in the twenty-first century, implies technological mediation. Concealed in this logic of stimulus, either mediatic or maternal, either too much or too little, either hot or cool, is the reduction of women (and other parents) to their role as a mother. In turn, a mother is reduced to a maternal function (a mom in Strecker's moment, or a mama, arguably, in ours), and that function reduced to the production of stimulus. This produces an inescapable problem wherein all that can ever be offered is too much or too little stimulus—precisely because stimulus is not the right conceptual frame for relationality: it is misogynistic in its description and in the outcomes and ramifications that this description will predict and observe. The persistence of mechanical and temperature metaphors for maternal care—far beyond their mid-century inception—indicates that though the devices, diagnoses, and metaphors shift, this underlying and punitive media theory of parenting remains the same. This is the other double bind of parenting.

4 SCREENING MOTHER, CODING BABY

On December 1, 2021, the state assembly in Wisconsin heard testimony on Bill 627, which was introduced two months prior to augment current laws that restrict parental rights in the case of incarceration. Proponents of the bill argued that it will provide children, many of whom are in the foster care system and in contact with Child Protective Services (CPS), some "much needed stability."[1] This stability is figured as being that which matters most to children, both psychologically and developmentally; what matters much less is the person upon whom that stability depends. A mother here can easily be exchanged for an other, so long as that other is stable. There is no notion that stability would be best ensured by producing it through the decarceration of parents. Instead, under this new law, parental rights can be terminated if the case meets three criteria: the child is already known to CPS, the parent is currently serving a term of four or more consecutive years, and—crucially—the parent is judged *likely* to be in prison for a "substantial period of the child's minority."[2] The bill was passed that day.

Wisconsin has the highest rate of incarcerated Black adults (one in thirty-six) in the country; in tandem the rate of women of any race incarcerated has grown sixty-six-fold in the last fifty years.[3] It is one state among many that uses algorithms in sentencing, which have been shown time and again to re-entrench the logics of racialized incarceration, routinely increasing sentences for Black defendants.[4] The state has long been a flash point for advancing computational control in the carceral, and setting policy for the nation. Whereas previously a judge had total control over sentencing within the federal minimums and guidelines, Wisconsin has used Northpoint's Correctional Offender Management Profiling for Alternative Sanctions (COMPAS) for the last five or so years.[5] COMPAS over-predicts the likelihood of all women to reoffend; it is also 77 percent more likely to declare Black defendants of any gender violent reoffenders. Recidivism is the ostensible motive for such algorithms, but, as many legal scholars, social scientists, and prison abolitionists have shown, this is far from the complete story. These algorithms are a tool of the *carceral imperative*: they are designed to make use of proxies for race and class in the markers they select for and are predicated on factors such as geographic region of residence, identity, and previous criminal record. As one element of the computation, these algorithms claim to draw on the dubious method of psychometrics via taking a psychological inventory—offered as questionnaires that are not administered by psychologists or psychiatrists. Sentencing algorithms (which do not feature machine learning, and are more or less just a formula with transparent variables) in part "compute" a different profile—a psychiatric one—a cosmetic way of cloaking and justifying the outcome, borrowing on an adjacent authority and discipline.[6] Sentencing algorithms and the psy-evidence that accompanies them aid in keeping control over "the bottom quartile" of Americans and drive the prison-for-profit model, what Jackie Wang calls "carceral capitalism."[7]

The history of maternal-child dislocation is crucial to the history of racial capitalism, long before the digital turn.[8] The destruction of kinship networks has subtended American capitalism for four centuries. As Mary Pat Brady writes,

> From its inception as English colonies to the present day, the United States has always relied on dismemberment, beginning with African, Irish, and English children spirited away, kidnapped from their parents by the Virginia Company to work its plantations. It has similarly relied on a practice of shredding affective relations to build and maintain a labor force: not only were indentured and enslaved peoples denied the opportunity to establish permanent kin networks or reliable affective structures, but so, too, were the men recruited from China and Japan in the nineteenth century, as were the men knitted into the Bracero program in the mid-twentieth century.[9]

Not only are children taken from their mothers but mothers are taken from their children. This double dislocation of mother and child has been part of the torn fabric of kinship in the United States, and carceral capitalism and its antecedents have upended American family life as they have done since its founding.

Beyond the state of Wisconsin, one in five Americans has had a parent incarcerated.[10] In Wisconsin, and in states with a similar admixture of algorithmic sentencing and restrictive parenting rights laws, children and parents will increasingly be severed from one another along racial and gendered lines, with Black mothers the most likely to have the harshest sentences and be flagged as likely to reoffend and thus ineligible to parent. If their children are already known to CPS, itself an organization that tends to separate Black parents from their children (as the crucial scholarship of Dorothy Roberts and Mical Raz has demonstrated), Black families are suddenly set to meet all three criteria as laid out in Bill 627.[11] The United States is only one of four countries worldwide to routinely separate incarcerated parents from their children; Bill 627 now extends this practice and further justifies it. Grounded in a notion of childhood *stability*, the bill newly determines who is a "fit" parent and thus who is eligible—or ineligible—for parental rights.

Stability as a watchword for healthy childhoods has a long history, one that emerges in the same period as the "mother-as-medium" discourse this book has traced. Stability is a softening of John Bowlby's "security." Where Bill 627 aims to provide "stability" for a child with an incarcerated parent, there is no provision for children born in prison. Some 8,000 pregnant people are in prison at any given time, and the most common health concern in women's prisons today is pregnancy.[12] It was and continues to be assumed that those who give birth in prison will not have autonomy over birth or the postpartum period, may give birth in shackles[13] or with armed guards present, and then endure a cruel and unusual punishment: having their infants taken away from them on the day they are born. Again, this is described by many psychologists and policy makers as for the good of the child, and in the name of childhood stability. This stability is assumed to be external and material rather than psychical, something the incarcerated parent cannot provide by virtue of being absented and withheld by the state.[14]

This insidious logic—that to produce stability, children must be torn from their parents, especially their mothers—is not new. As historians Caitlin Rosenthal, Walter Johnson, Jennifer Morgan, and Brenda Stevenson, among other scholars, have shown, actuarial management of the family and of the mother and child bond was central to the practices of the plantation, including the near-ubiquitous practice of family separation in the antebellum context, as well as carceral practices in the Jim Crow era.[15] Joy James terms these women "Captive Maternals"—drawing a genealogy across the last 300 years to argue that the United States' "longest war is with its domestic target: enslaved or captive black women, a war that dates back to the Commonwealth of Virginia's 1658 attempts to (re)enslave Elizabeth Key, one of the first Captive Maternals to have her battles enter public record."[16] The very notion of fit mothering has its modern basis within Progressive-era eugenics, bolstered by statistical models centered on heredity as well as the co-emergence of social work and the expansion of the Child Welfare Bureau at the behest of the newly codified social worker (into what would later become CPS).[17] As Dorothy Roberts argues, starting with this agency, social work, via its reach in CPS, has targeted the Black family, creating "shattered bonds" for a century.[18] Or as Joy James writes, "Centuries later, Black Captive Maternals remain disproportionately disciplined, denigrated, and consumed for the greater democracy."[19] Control over the mother incarcerated is a remainder, a continuation of transnational enslavement protocols of control.

Management of the incarcerated "Captive Maternal" has partially centered on her children. As Estelle B. Freedman shows, maternalism has long been linked to notions of incarcerated women's reform and recidivism, starting in the 1840s with the gendered segregation of prisons.[20] That is, to make fit citizens, women first had to be reeducated as mothers. In parallel, as prisons become segregated according to gender, the practice of keeping mothers and their children together came under regulation; there were fewer and fewer women's prisons—until 1980—that allowed incarcerated mothers to mother much at all.[21] Put simply: mothering has been linked both positively and negatively to redemption in the penal context since the late nineteenth century and has been a focus of carceral control. This chapter tells the story of how the psy-ences reconfigured mother as a medial environment, as a site of security, within the securitized environment of the US prison.

Across the last eighty years or so, notions of psychological developmental outcomes in children and the status of parenting in prison have become intertwined—the psychological apparatus and the carceral apparatus have routinely commingled both experimentally (psy-experiments in prison) and legally (the language of attachment deployed for and against parental rights). As Malcolm X testifies, the destruction of his family was due not to the psy-ences nor to the carceral system alone, but both at once:

> I truly believe that if ever a state social agency destroyed a family, it destroyed ours. We wanted and tried to stay together. Our home didn't have to be destroyed. *But the Welfare, the courts, and their doctor, gave us the one-two-three punch.* And ours was not the only case of this kind. I knew I wouldn't be back to see *my mother again because it could make me a very vicious and dangerous person—knowing how they had looked at us as numbers and as a case in their book*, not as human beings. [emphasis added][22]

When we think about the human sciences at mid-century, we may be aware of a widespread focus on delinquency in children, one at the heart of Cold War social science. I complicate this account by arguing that theories of mother-child bonding were theorized and enacted in prison and its courts via its doctors across the twentieth century United States into our present moment while, for the psy-ences, the controlled environment rendered the prison a suitable, temporary laboratory, where the data of mother-infant bonding was exploited. These subsequent theories of mother-infant bonding—how mother regulates infant, how she becomes an environment unto herself—turn mother-as-medium into the basis of long-term prediction, and reconfigure methods for familial control. That mother might not be the right medium of care became a question raised in the mind sciences, and then answered in the carceral system.

I open in our present and extend backward from the digital to show how policies that undermine mother-child relations are co-produced by the very theories of maternal presence, absence, and environment advanced at mid-century. Theories of maternal effects, attachment, and care at mid-century engendered predictive logics of fit mothering (in the next chapter, we will see how these maternal effects became tied up in media effects). In this chapter, I turn to one of the momentary laboratories where a longer-term maternal effects theory was evinced: the women's prison and the mother-infant nursery program therein.

If, at mid-century, much of the attention of the human scientists interested in the family was directed to how mothers impacted infants *immediately*, theories of deprivation and attachment also looked at the consequences of qualities of mothering long-term. As US carceral policy meets digital technologies, as in the case of Wisconsin, we can see the antecedents of our contemporary predictive tools in these earlier, analog logics. The roots of this "digital logic" take shape in so-called analog forms (in architecture, in psychological research, in the segregation of races and genders and classes).[23]

As we have seen across this book—from "Little Albert" to John Gray Jr. in the Air Crib to Leo Kanner's autistic subjects at Johns Hopkins and "Joey" in Bettelheim's Sonia Shankman Orthogenic School—the history of childcare and the history of child experiments sit uncomfortably close together. Almost always, these children are contained, separated from their parents, even if momentarily, and the maternal environment is exchanged for a material one. Rather than emerging solely in a lab or in the clinic (although these locations are crucial too), I argue that these predictive theories of maternal care can be productively considered when linked to Viennese psychoanalyst René Spitz (1887–1974) and his work in an American women's prison. Bernard Geoghegan argues that in recent years historians and media theorists have found that the clinic, the zoo, and other momentary labs that served as strange experimental sites for data-gathering are decisive for present-day new media.[24] Much of this book situates child-rearing where it meets communication theory. These stories of science and technology—and particular kinds of mother-child dyads—may be termed pathological, but they evince a genre of white middle-class medically privileged mothers and their offspring as setting the paradoxical pathological norm, as we saw in the previous chapter. In this chapter, we will see that maternal environments were co-present with the environment of the prison, which allowed researchers to look at the endurance of maternal love in the infant during psychical experimentation occurring therein. This has largely been left aside by historians, yet it is crucial for those of us doing longer histories of technology and psychology that culminate in our contemporary moment and its regimes of capture and control, allowing us to get outside the false logic of "digital exploitation" as specifically digital. To do so, we need to move beyond technological determinism to the history of these forms of familial violence that are as rich and layered as "race" and "gender" under the US state apparatus.

DEPRIVATION THEORY

"Attachment theory" is a misnomer. It is a theory of deprivation. Across its longer history, theories of maternal-child bonding have fluctuated as to the scene they describe: mother and child apart, mother and child together, or a kind of psychic distance that crosses both togetherness and separation. In brief, based on nearly a century of infant-mother observation substantiating the impacts of mother-infant separation and, later, to distinguish types of attachment, attachment theory relies on the longstanding tradition of mothers, and then psychologists and pediatricians, watching children—in their homes, through a mirror, or, later, on videotape, as they respond to stimulus or its dearth. The location of these observations is eventually assumed by the clinic, the laboratory, the early childhood program at the university. The subject of observation is almost exclusively the child; the clinicians performing their analyses, however disparate, have been termed "baby watchers"—crucially *not* mother watchers, however much they may assess them too. As attachment theory has become domesticated, the assumed mother is—if present—one who lives with her baby at home. However, the science was born from entirely other domestic conditions.

Starting in the 1930s, the practice of child psychoanalysis grew in recognition, if not in prestige, in Central Europe and England. What emerged across the era's consideration of mother-infant relationships and child outcomes was, as I have argued throughout this book, a taxonomy that centered on the dosage and quality of attachment, presence, absence, and affect. Nicolas Rose describes this as an obsession with "the minutiae of mothering," which he dates to the postwar era.[25] However, we can see this focus on micro-mothering presaged in the early thinkers of mother-child bonding that proceeded attachment theory, including British object relations theorists such as Melanie Klein, as well as ego psychologists such as Anna Freud at work in the United Kingdom on caring for children orphaned (even if for some separation was temporary) during World War II.[26] All of these vastly different conceptions of the mother-child relationship take as their central concern that how a mother and baby get on, how they bond and attach, will have ongoing effects on the child's development. All of these theories are, at their base, predictive. All of them argue that a faulty relationship between mother and child can have dire consequences.[27]

The qualities of this early relationship or, in the parlance of John Bowlby, "attachment," stipulate *security* or its absence (as theorized particularly in the work of Bowlby and Mary Ainsworth, starting with the latter's dissertation on the topic). When a child is not "securely attached," a host of specific negative outcomes are to be expected. While Bowlby began to formulate the earliest versions of attachment theory in the 1950s through children he observed during hospitalization and institutionalization, his earlier work a decade prior focused on delinquency in children—as evidenced by his landmark 1944 paper, "Forty-Four Juvenile Thieves: Their Characters and Home Lives."[28] The figure of the delinquent child held great fascination for many mind scientists in this period—from Karl Menninger in the United States to D. W. Winnicott working nearby. In his work on the delinquent child, Bowlby combined psychoanalytically inflected case studies with statistical tests to provide a composite picture of the delinquent teenager; this actuarial data is inflected with the if-then logics inherent to algorithms. The research gazed backward from the present to infancy and childhood in a proto-algorithmic fashion to determine the cause of delinquency, using an admixture of statistical portrait of the cohort, micro-case history, and observation to tease out psychotic and "affectionless" character in the latency period.[29]

While Bowlby's attachment theory has become, as historian Marga Vicedo argues, the most well-known child development theory from this period, with huge implications for policy worldwide (including, as we saw in chapter one, the shuttering of nursery programs in the UK), René Spitz's work was also foundational to arguing on the world's stage that mothers should be with their children—no exceptions. As is revealed in Bowlby's correspondence with D. W. Winnicott, Bowlby's work in Britain was later used to shut down day nurseries. Winnicott wrote to Bowlby, "It has come up against quite a lot of people who are worried about the way your work has been used by those who want to close down day nurseries. . . . You will probably agree with me, and I would very much like to be able to say in public discussion that you agree, that there is a deplorable shortage of Day Nurseries accommodation. . . . This is a vital problem and I am afraid that at the moment your having been quoted in connection with the closing down of the Day Nurseries is doing harm to the very valuable tendency of your argument. I wonder if there is anything you can do about this."[30] While Bowlby was highly critical of Spitz, in part because Bowlby discarded the psychoanalytic account of mother-baby altogether, making the bond innate and primary, Spitz had already pointed the way to Bowlby's maternal deprivation hypothesis by observing the child at a much earlier stage.[31]

Spitz's work focused on the resulting state when an infant is separated from their mother for an extended duration. He investigated separation outcomes in the United States in hospitals, especially at Mount Sinai, where he researched anaclitic depression (the state resulting from the partial loss of a loved one) and "hospitalism" (resulting from the total loss of a loved one). Spitz codified this latter term to mean the nonfatal "wasting away" of the infant when deprived of contact with the lost person (read: mother) for more than five months (distinguishing duration of separation will be key vis-à-vis parental sentencing; the longer the sentence the more likely, on the grounds of attachment theory, parental rights will be terminated).[32] As historian Ellen Herman suggests, Spitz's work inflected policies central to adoption, noting that, when it came to connecting with the mother, "Better late than never" was not an option. Spitz went on to continue this work in orphanages in Mexico and South America and, crucially, in a prison nursery, where mother-infant pairs would be together for a time and then separated.[33] These institutions became momentary laboratories—the clinic and prison converging in the site and act of surveillance. The very captivity of the mother in prison allowed the investigator to focus on and isolate the impacts of maternal love, for mother was not free to be elsewhere.

Spitz's research in the prison was also centered on the infant's maternal environment and its influence on their development. Spitz was careful to say, rearticulate, and confirm that the *environment* here is the maternal one—not the physical space a child is in, and not even the father, if there is one.[34] He conflates maternal love and attention with an enveloping medium. Spitz is interested in the total environment of maternal love—or its contingency and absence. To substantiate his findings, much like Bowlby's work in the 1940s on delinquent teens, Spitz, as Rachel Weitzenkorn argues, yoked the empirical and the psychoanalytic. To do so, he turned not only to statistics, as Bowlby had, but to film in order to corroborate the batteries of tests to which he routinely subjected the mothers and infants.[35] As part of the Psychoanalytic Research Project on Problems of Infancy, Spitz made a number of evidentiary documentaries (with Katherine Wolf as cinematographer), the most circulated entitled *Grief: A Peril in Infancy* (1947), which records the psychic life of infants in a foundling home.[36] The silent film, which makes use of intertitles, proceeds in three parts. Part one, centered on a Black infant called "Jane," commences with the intertitle: "I. The Baby Loses Her Mother." Spitz is shown bringing his face all the way down to the baby, visiting her, trying to comfort her. In the second part, she, and other children (who remain unnamed), become listless, wasting away. The final part begins with the triumphant: "III. THE CURE—Give the Mother Back to the Baby." The children are seen as immediately revitalized. Even though still in care, presumably mother is visiting. Infants climb out of their cribs, are spirited, and are cured through this reunion.

This was Spitz's basic program: mother love was a cure for motherlessness. Conducting his research then required investigating different arrangements of the comings and goings of mothers. To do so, and across his experiments, Spitz observed infants in a wide range of environments, including a control group of middle-class children in private homes, and two contrastive groups of infants—in prisons and in orphanages—to provide the basis of comparative studies.[37]

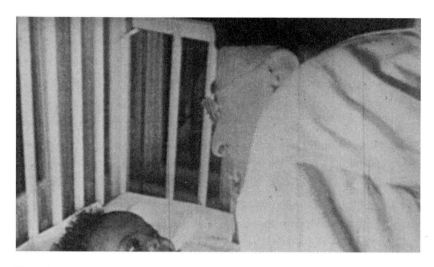

Figure 4.1
Still from *Grief: A Peril in Infancy* (1947), from part III: "Give the Mother Back to the Baby." The Cummings Center for the History of Psychology, University of Akron.

The prison nursery is held up as the example of *better* emotional care and outcomes; mothers are allowed routine contact with their children and the infants are cared for by a full staff.[38] Spitz visited the nursery for five years, where he and his team tested the infants with "Buhler-Hetzer baby tests, alternating the sex of examiners... and gave Rorschach and Szonzi tests to a group of mothers in the nursery." They then "took motion picture of certain modes of behavior which were particularly characteristic of, or deviant from, the general behavior of the group at a particular age." The filming "allowed the viewer to watch minute details and do that with comparative ease"—something Spitz remarks is "impossible in the actual observation of these infants."[39]

Grief and other similar films in this period, as Katie Joice shows, took the focus on the pathogenic mother and turned it to account for the infant, even as the emphasis on the etiological fault of the mother remains through the 1980s and beyond.[40] Felix Rietmann argues that Spitz's work spans this shift: "Spitz and Wolf's work was a forerunner of a host of psychoanalytically informed observational and longitudinal studies of infant development that flourished in the 1950s and 60s. Almost all of these studies sought to integrate psychoanalytic reasoning with empirical and sociobiological data."[41]

Yet in Spitz's published research focusing on these comparative studies, the prison under observation is never named directly, nor is the mother; bonding and care are made decontextualized evidence, even in her very exclusion. In Lisa Cartwright's terms, Spitz becomes a "moral spectator" in keeping with child analysts' work in this period: "Psychoanalysts broke with professional protocols to actively engage with their subjects in an exchange of looks and physical interaction to intervene in the crises they witnessed."[42] As we can see in *Grief*, as Weitzenkorn argues, the researcher (Spitz in this case) takes on the role of the mother; it is he who elicits the stimulus.[43] He literalizes the emergent idea that the therapist is also a maternal proxy, and can do the work of standing in. He shows just how limited that view is when it comes to infants. On film, the baby takes center stage (or is the subject of a close-up, which, as Mary Ann Doane tells us, betrays intimacy with the viewer).[44] Yet even as the researcher replaces the mother, she is the absent presence under consideration. Her child is predictive evidence for the result of her care.

THE CURE FOR MOTHERS

It is no secret that Spitz worked within a prison nursery, and yet scholars have neither foregrounded the fact of working within a prison nor paid specific attention to the actual institution he was using; the fact that the field laboratory is a prison nursery, let alone this one, has remained a footnote. But tucked away in Spitz's archive is a note that the prison that served as an ad hoc laboratory is the Westfield State Farm Reformatory in Bedford, New York.[45] By making Spitz's work at Westfield State central to the conversation of infant observation and its hundred-year-long history, we can trace theories of mother-baby interactions' long-standing relationship to incarcerated mothers as a primary scene of separation and deprivation and identify a central node in the theorization of what would later be called attachment theory and its attendant norms of care and reproduction.

Spitz was emphatically pro-maternal, but in the prison he interacts scientifically rather than historically with the carceral's long-standing relationship to mothering. The relationship between reformability and the maternal is always moving in two directions at once, tautologically: mothering can aid

Figure 4.2
Women prisoners in the dining room, New York State Reformatory for Women, Bedford, New York. University of Warwick, Modern Records Centre, Photographing Prison Life in the 20th Century collection.

reformation, but the mother is the reformable subject who is not to mother unless she is reformed. In omitting the prison as context for this theorization, we perpetuate Spitz's mistake; obscuring the prison as the hidden foundation for his maternal environment. This under-acknowledgment is precisely part of Spitz's understanding of the mother and infant relationship. Therefore, mother-baby interaction's precedent and its afterlife are part of what Tonia Sutherland calls the "carceral archive."[46] By locating it as such, we can follow a different trajectory of maternal surveillance as it relates to these theories, one that comes from the carceral and stays within its ever-expanding reach as, in our present, maternal "fitness" is predicated on *stability* and *attachment*. In parallel with these other mediated forms of watching, this one is mediated by the co-constitutive apparatus of psychiatry, state service, and prison.[47]

Westfield State Farm Reformatory has its own particular history as regards both the psy-disciplines and, following Joy James, the "Captive Maternal": it is home to the oldest US prison nursery, which opened in 1901, in a period when, as scholars of the women's prison, like Estelle B. Freedman, show, mothering was central to notions of women's criminal rehabilitation. In the Progressive era, prisons reprised antebellum logics of family separation on the plantation.[48] Babies were largely taken from their mothers, and rehabilitation became a science to be studied. Reformatories established posts for psychologists and social scientists to study the problems of "the fallen woman," with incarcerated women becoming raw data for new diagnoses (namely, psychosis) as well as providing up-close examples of "unfit mothering."[49] Westfield State has additionally been the site of other psychological and experimental research, including a Rockefeller Foundation-sponsored laboratory to evaluate and develop new treatments for "feeblemindedness," and, as Nancy Campbell has shown, in the period directly following Spitz's work, the prison also housed a section for drug users, who were increasingly criminalized in this period.[50]

Regarding mother-infant dyads, women's prisons in the north of the United States, where Spitz was working, had divergent policies by this time. Some separated mothers from their children as matter of course, while others allowed for some pregnant "fallen women" and their children to remain linked, in accordance with notions of reform via maternalism.[51] Westfield State fell into this latter category. As Ilse Denisse Catalan shows, the project of reforming the largely young women's population—either already mothers or pregnant—was to *train* women into becoming mothers. This was in part because they were assumed to have had unfit mothers themselves—the sign of which was the fact of their imprisonment.[52] What Spitz captured and theorized were the bonding experiences of mothers under the control of the state, who were already under surveillance from the increasingly powerful psy-disciplines. Unlike some of his analytic counterparts, like D. W. Winnicott (who, as we saw in chapter 2, argued that good mothering cannot be taught and thus unfit mothers are not reformable), Spitz thought of all mothering as essential—and his psychoanalytic findings could thus be instrumentalized to keep mothers and babies together as a general case.

But the formation was not one born out of a generality: Spitz was working within a single prison, one that had a long-standing nursery, where this reunion and separation were repeated again and again within the confines of the prison itself.[53]

While the archival documents of the nursery at Westfield State in Spitz's exact moment of experimentation and filming were destroyed in a flood, nursery day books from the next consecutive years allow us a glimpse into the highly regimented comings and goings Spitz was studying (although sometimes it is erroneously reported that Spitz studied penitentiary mothers who had access to their babies all the time; this is evidently not the case). In the 1950s and 1960s, as documents from the Westfield State nursery programs suggest, other kinds of maternal-infant reunion, separation, attachment, and instability, as well as child-viewing and maternal-surveillance, were possible (or impossible).[54] Babies were surveilled medically on a regimented schedule: temperature checks twice daily and weighing twice per week, recalling recently bygone models of scientific mothering. Sick babies were routinely isolated from their mothers—even though this practice flew in the face of the research being conducted in the same unit.[55] How and when mothers and babies were allowed to see each other was completely controlled. Outside time was either limited or nonexistent.[56] Inside, mothers could see their babies with written permission, but sometimes only through a window. Other mothers snuck visits to their babies; if caught, they were written up.[57] Here, we can see another tradition of observing infants with their mothers running in parallel to the state-sanctioned psychological experiment, yet one still managed by the state. In seeking to document his experiments for his own research program, Spitz mimicked the practices of surveillance already endemic in the nursery.

Nevertheless, this work in a prison nursery has been by and large submerged in the longer history of maternal theories and its attempts to locate and quantify qualities of what will later be called "attachment" (and deprivation) as they relate to pathological and delinquent states in children. The story of attachment is often told as one of war: born in the transient conditions created by World War II, with orphans providing new data, or via the Cold War science that followed. It is often told as one tied up in the social panics over the delinquent child. I add to this account that attachment theories were escorted by an implicit codification of mother-as-medium, and then by psychologists joining with state officials to regulate mothers like a medium as they envelop their children. Here, we can see Spitz was not just focused on the orphan as scientific type, with which he is most normatively associated. Theories of maternal-infant love are inseparable from their contexts. But these laboratories, the evidence gathered therein, and the theories generated from that evidence, departed the carceral context and traveled back to the nursery. As Ellen Herman has it, describing the swath of research conducted at mid-century on "attachment, "[i]nvestigations of abnormal conditions in abnormal environments produced knowledge about normality that applied to all children."[58] The early focus in mother-infant interaction theory on predicting the emergence of the delinquent child—one who does not learn to attach early and often—is often told as a story where mother was abnormally absent. Here we can see it also has its roots in the observation of American incarcerated mothers undergoing managed separation from and reunion with their children.

Later, as attachment theory gets codified (more about this below), it concerns the prediction of bad developmental outcomes for children. As I have argued across this book, these outcomes are always etiologically traced to the mother on the grounds that she has provided too much or too little stimulus, notions of which are tied up in conceptions of mother-as-medium during and after the Progressive era.[59] Whereas this focus on childhood delinquency and bad mothers has had a long life outside the carceral context, here I seek to resituate the prison as a site of psychological experimentation on mothers and infants. This normalizing, diagnostic imagining of the mid-century mother-infant relationship continues to inflect not just the observation of mothers, but the management of their bonds from the carceral clinic and beyond.

CAPTURING THE CONTROLLED

Spitz was not alone in turning the science of bonding into actuarial data. His project anchors a genealogy of baby watchers who, by the mid- to late twentieth century, took up other forms of documentary observation, moving from film in the 1920s (Arnold Gesell) to tape (Mary Ainsworth) and eventually to the digital image mobilized by clinicians working in the twenty-first century.[60] As scholars such as Felix Rietmann and Lisa Cartwright have shown, infant observation has generated norms for diagnosis and screening of a variety of maternal and infant pathologies.[61] But these theories did not stay only in their clinical (and clinical-carceral) contexts; they then bled into mainstream parenting discourses across the second half of the twentieth century.[62] As Rietmann argues, when child observation met cybernetics (sometimes in ethnopsychoanalysis), how infants were perceived changed, such that they were "no longer considered victims of their environments but endowed with remarkable social and communicative capacities. It was no easy task to show these capacities," for "it depended on a considerable recalibration of scientific and medical vision."[63] While I am arguing that this recalibration was already taking place a decade earlier, by the 1970s, this hybrid of psychoanalysis and experimental psychology had fully adopted new empirical modes presaged by Spitz. From him, it inherited the use of moving image—now the video camera—and the coding of maternal and baby interaction it made possible under a new media regime, as that interaction was now pausable and rewindable. It could, in essence, become coded—both as data and as descriptive prediction. This long arc of documentary attachment theory starting with Spitz anchors a prehistory of facial recognition, data coding, and diagnostic algorithms in the relationship between mother and child. Although media technologies had been previously included in documenting the mental life of the infant, from early still photography through films by psychoanalysts and auteurs alike, starting in the late 1960s, the new medium of videotape and its evidentiary capacity greatly enhanced the data-gathering capabilities of clinical psychologists (as well as amateurs at home). This allowed for the aggregation of attachment data—moving it from theory to science.

Most famously, psychologist Mary Ainsworth turned to the use of videotape to code the infant's attachment style in relation to a maternal stimulus. To do so, Ainsworth made use of videotape for her "Strange Situation" classification (1969). The Strange Situation is an eight-part test in which a nine- to eighteen-month-old child, mother, and stranger are brought together and separated in a series of entries and exits to an observable room:

1. Mother, baby, and experimenter (lasts less than one minute).
2. Mother and baby alone (three minutes).
3. A stranger joins the mother and infant (three minutes).
4. Mother leaves baby and stranger alone (three minutes).
5. Mother returns and stranger leaves (three minutes).
6. Mother leaves; infant left completely alone (three minutes).
7. Stranger returns (three minutes).
8. Mother returns and stranger leaves (three minutes).

The entire undertaking or test is filmed (and the test continues to be given; see below). Of primary interest to attachment researchers is what happens during the two maternal returns. Children are then scored on a scale of 1–7 on four criteria:

1. Proximity and contact seeking
2. Contact maintaining
3. Avoidance of proximity and contact
4. Resistance to contact and comforting.

These scores are predicated on micro-observations, conducted every fifteen seconds throughout the duration of the test.[64]

The Strange Situation reprises and rhymes with the choreographed separations of regulated access and care of infants Spitz witnessed. Unlike his earlier films, the detail-oriented practice of coding the faces by time and quantifying the behaviors of videotaped mothers and children did more than just render these phenomena more visible. New diagnostic categories and their criteria were developed out of this closer, pausable, rewindable, and reviewable form of looking. While much of the literature has focused on the ostensible object of this clinical surveillance, the child, videography in the clinic allowed for the eventual coding and typing and pathologizing of "bad" mothers and "good" mothers, "sick" mothers and "well" mothers, "secure" infants and "anxious" infants (as in the studies of Ainsworth and Bell; Bebe; and Tronick), but also allowed for more focused research on the child. From this constellation of experimental psychological attempts to norm secure and pathological attachment, we can trace across media regimes a prehistory of emotion recognition and diagnostic algorithms in the relationship between mother and child when they become scalable data. Mothers and their children have been central to elaborating these practices of prediction based on micro-expressions and nonverbal actions, as film historian Seth Barry Watter has shown.[65] Across these experiments, the mother becomes increasingly beside the point even as she remains the object of scrutiny: she was the stimulus and one frequently deployed off-screen. The diagnostic focus is all about the baby. Again, theories of attachment were invested in the here and now to predict a then and there; they were predicated on a pathological mother, begetting a pathological child, or a good mother, begetting a securely attached child. These norms and their evidentiary regimes further the if-then of predictive behavior at the core of attachment theory, merging with carceral forms of algorithmic behavioral prediction.

ACTUARIAL RECIDIVISM, ACTUARIAL ATTACHMENT

As attachment theory continued to screen and code the infant and their maternal stimulus, its premises and visual evidence begin to find their way into the courtroom.[66] In the 1970s and 1980s, the role of attachment theory in legal frameworks had a dual purpose. On the one hand, the theory was deployed to uphold decisions around revoking custody—attachment theory's predictive science become a state technology of control. This is at work in Wisconsin's Bill 627 some fifty years later: a strategic fantasy of childhood stability in the future leads the court to argue for revoking parental rights in the present. On the other hand, these same mothers turned attachment theory to account by using its logics to argue for the legal right to bonding time with their children. This legal right was enshrined unevenly in prison-run mother-infant programs, which, while extant across the twentieth century, were expanded under logics of rehabilitation. It is this flexibility within attachment theory—even as it becomes an ever more quantifying science—that allows it to have two parallel trajectories from the 1970s to present.

Attachment theory commingled with extant regimes of actuarial information, which had long been part of pre-trial, sentencing, and parole.[67] "Captive Maternals" were increasingly incarcerated beginning in the 1970s. The US legal system shifted its focus from discretionary sentencing to federal mandatory minimum sentences for drug-related offenses, culminating in the 1986 Anti-Drug Abuse Act.[68] As a result of these laws, and the move to include statistical, proto-algorithmic sentencing, the population of women's prisons skyrocketed, nearly doubling in just five years, from 42,176 (in 1985) to 81,023 (in 1990).[69] In tandem, statistical models for prediction, once used to gauge the likelihood of recidivism during parole hearings, moved into the primary legal scene of sentencing (although they do not in all instances lead to increased sentencing).[70] In 1984, the Sentencing Reform Act begot a federal instrument that set a statistical model in place for determining sentences.[71] The prison population grew exponentially as these mandatory sentences were intermixed with algorithms that produced new norms around sentence duration.

Beyond increased incarceration, and who is subjected to it, the *length* of a prison sentence matters for the chance to parent one's children after imprisonment. While long and short sentence durations (including pre-trial detentions) each have their unique impacts on individuals and families, sentence duration is key for family systems and those parenting from within the carceral system, where 58 percent of parents do not see their children for the duration of their sentences. For the parents in question, the longer the sentence, the less likely it is that they will be granted access to their children.[72] Visits with children and "active parenting" have long been crucial to retaining custody since the reorganization of Child Protective Services across the 1960s and 1970s.[73] As carceral computing further entrenched and exacerbated over-sentencing and mandatory minimums in the Clinton and post-Clinton era, sentence duration became actively intertwined with legal definitions of "fit mothering." In the context of these state policies, the focus remained on the child being placed into a family—any family—over the rights of the incarcerated birth parent.

FIGHTING TO BOND

For women who are incarcerated while pregnant, or who become pregnant during their sentence, the right to parent within prison meets these same legal challenges, commencing prenatally. Beginning in the 1970s, incarcerated mothers fought legally for their right to parent their children *while* doing time. After much agitation from formerly and currently incarcerated mothers, activists, and lawyers, prison nurseries or mother-infant programs—where mothers and their infants are allowed to stay together—were expanded to allow for longer bonding periods and were more frequently mandated and available.

The first case of such a fight was that of Terry Jean Moore, which received some mainstream notice.[74] Moore had been sentenced in 1977 to over seven years for a five-dollar robbery—a sentence that was unusually harsh. In 1979, Moore was pregnant and began to fight to keep her pregnancy, and then her baby.[75] To do so, Moore made use of a little-known statute, one never utilized in her state of Florida, that allowed parents to keep their babies with them for the first eighteen months of the baby's life.[76] Moore's case brought the statute to light, and the attention on it quickly left it precarious. Prison superintendent William E. Booth immediately moved to have the law removed, arguing, "This is no place for a baby. . . . A kid ought to grow up in a home."[77] This is the fulcrum on which Moore's case and others like hers lie: Where ought a kid grow up, what constitutes a home, and what status does the maternal environment have? To advocate for Moore, a white woman, her lawyers argued that in essence they agreed with the superintendent, but with a twist: the mother (in this case Moore) *was* the home and the best one—the total environment of love, as Spitz had insisted. The prison walls did not matter. Because the statute made it illegal to separate mother and child without cause, Moore's opponents had to name incarceration itself as a criterion for child separation, a burden they were not able to meet. Moore won her case. She and her daughter Precious were allowed to stay together in a prison nursery, with thirty minutes per day outside in the yard, and with Moore receiving childcare in the form of two hours away from her child when prison guards took care of her child in her stead.[78]

A legal landmark, this case demonstrates how slippery attachment theory is in the carceral setting—how inherent its interpretive flexibility is, how easy it is to bend toward or against any particular aim. Moore's lawyer, Jacqueline Steinberg, gestured at attachment and its lack by claiming Moore's destiny had been pre-written because Moore herself was a product of the foster care system.[79] Speaking to the press after winning her lawsuit, Moore articulated the notion that being with her daughter not only gave her a connection to her child but a connection to the notion of a post-incarceration future. She hoped her story would do the same for others and serve as stark warning: "no matter how small the crime, they could waste their lives away behind prison walls."[80] The notion that keeping mother and baby together serves as a deterrent to crime and to recidivism reappeared again and again in the cases that followed in the next decades.

After Moore won her lawsuit, several American prisons began to elaborate means by which incarcerated mothers and their newborn infants might stay together for the short duration of a year or eighteen months; they were and are much more accessible to white women. Central to the push for these programs under the law was attachment theory, and its language of bonding, already racialized. Using its prescriptive predictions, lawyers were able to argue that mother and child belong together—that the maternal environment is paramount regardless of the physical setting.

In the 1990s, after waves of lawsuits following Terry Moore's case, prison nurseries became more common but far from ubiquitous. In the United States, there are now twenty states that offer such programs, and this number was reached when attachment passed from theory and experiment into an accepted state that underpinned good mothering, one as mainstream as any other even if its intensified and quasi-related form—the style of attachment parenting—remained on the fringe.

The right to mother, let alone intensively, is highly contested in this period, especially after the passage of the 1997 Adoption and Safe Families Act under Clinton.[81] This act changed the temporality of the termination of parental rights by mandating children in foster homes be available for adoption and the stability it could provide, uncompromised by incarceration; thus, much like Wisconsin's recent Bill 627, parents serving sentences longer than fifteen months were likely to have their rights terminated.[82] In turn, attachment as a logic was used to argue *for* the expansion of incarcerated mothers' rights. As I have shown, attachment arguments can serve opposing ends: *against* the incarcerated mother on the understanding that she is unfit for having "landed herself behind bars" or *for* the incarcerated mother as the proper, irreplaceable, and total environment of love.

In other court cases from this time, the law was mobilized *not* on behalf of progressive incarceration but instead on behalf of a conception of the child's future. As one example, in 2008, the supreme court of Pennsylvania held that a mother's parental rights should be terminated on the grounds that she was repeatedly incarcerated and thus *had not* bonded with a child—and that in spite of engagement with fifty-two weeks of parenting programming while incarcerated, she was in essence an unfit mother.[83] The ruling at once upheld the idea of a maternal environment and decoupled it from a specific mother. Similarly, in 2009, in the case of *Leeper v. Leeper* heard by the court of appeal in Louisiana, the court ruled that the fact of incarceration of the mother meant that the mother's parental rights were terminated in favor of the father—again using the language of attachment (and contact between mother and child was restricted to the epistolary).[84] As recently as 2021, in Tennessee at the court of appeal, a mother had her parental rights terminated despite being eligible that year for parole. The court decided not only that she was an unfit mother due to her incarceration history, but that her children had *bonded* with their foster parents—giving the foster parents custody under the language of attachment theory.[85] What these three cases share, beyond outcome, is that they all show the mother as meeting specific criteria for the loss of her parental rights—one of which is a likelihood of instability (of the child via the mother) in the future. This calculus is the predictive engine of attachment theory. Because of attachment theory's focus on the maternal environment, the term becomes, in a carceral and legal setting, one that can migrate from the mother's body *because* of her physical whereabouts to a new location/situation (whether the adoptive family, the foster home, or the unincarcerated parent). Put another way, when setting is fixed and immutable (the prison), the theoretical maternal environment can be moved to a new location or person(s).

It is not just the language but the documentary impulses of attachment theory that made their way into the courts. As one further example, in a 2020 case, *A.S. v. N.S.* in the state of New York, a mother was granted only supervisory visits to her newborn in part because she had previously been arrested; this hearing was subsequent to a filing by the father the previous year, when the baby was as yet unborn.[86] Several pediatric experts testified in the case, and reviewed videotapes that the father had taken of the mother without her consent—and which she had not been privy to. Despite being collected in this manner, the tapes were entered as evidence—and given scientific status because of their review by the child's pediatrician. Most specifically, they were used as evidence of the attachment between mother and child—as well as the mother's behavior. The following is the court's representation of the diagnoses of the pediatrician:

> The baby would prefer the mother's smell to the baby nurse's smell or anyone else so the attachment bond would not form with the nurse. She opined that while the videos she viewed (Court Exh. 1) demonstrated disturbing and emotionally deregulated behavior by the Mother, there were also parts that were emotionally nurturing parts, such as when the Mother said "I love you. I want you to eat." She opined that if the Mother has support and safeguards are put in place, having her be the primary caregiver would give the baby an advantage for the first 3, preferably 6 months, of his life, particularly given this child is at a disadvantage given the divorcing parents, and divorce is classified as an adverse childhood experience.[87]

The husband has in essence captured an accidental and amateur version of Ainsworth's "Strange Situation" test (see above). We can see the calculus of attachment theory at work here in the record—the tension between what is good for the baby and what might be bad, and how to legislate against negative outcomes. We can see equivocation between the two poles of attachment theory. The fact that the mother's emotional dysregulation may be compounded by the legal intrusion and mandated supervision as she breastfeeds, for example, is never mentioned. In this extreme example, we can also see the migration to the domestic, enacted by the husband, of the surveillance paradigm inherent in attachment theory's turn to videography and its subsequent use in and by the court.

ONCE A LABORATORY, ALWAYS A LABORATORY

Today, the exemplary infant program is housed in Bedford Hills Correctional Facility, in Westchester County, New York. This program and others like it display an uncanny admixture of the logics of attachment theory—itself a combination of experimental psychology and psychoanalysis—and carceral imperative. They bring the logics exposed in this chapter into contemporary expression.

It is perhaps no accident that the program is at Bedford Hills, which was once Westfield State Farm, where Spitz conducted his research in 1945. The program has been operational since 1901, when Westfield State first opened its doors, and has since informed the national conversation about incarcerated mothers and the right to parent while in prison. Yet despite its history of psychiatric oppression, starting at mid-century, the prison is the site of a long tradition of radical care coming from within—prisoners famously staged an uprising after the assault of an incarcerated woman in 1974; systems of peer care and AIDS counseling were developed by incarcerated women for incarcerated women; and writing groups hosted by activists and scholars inside and outside the prison paved the way for contemporary prison education programs.[88]

In the 1980s, the prison went through an additional period of reform, just as the women's prison population began to expand. Some of these progressive shifts occurred within the configuration of the prison nursery. Although the original mother-infant program was highly controlled and non-cohabitory, it became cohabitory through a series of moves to make the prison more progressive: a warden with progressive (though far from abolitionist) politics was made head just as an influx of progressive volunteers guided by Sister Elaine Roulet took over the mother-infant program (when she died, it was erroneously reported that she had helped create the program, though it had been longstanding). Most importantly, agitation from within by "the women in green," as the incarcerated mothers are called, continued. The result: the program shifted toward preserving maternal and infant bonding.

The program is presently run by volunteers, including incarcerated women and nuns, as well as psychiatrists, social workers, and correction

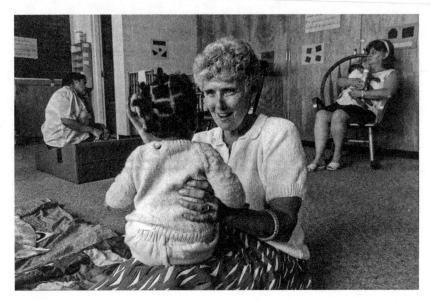

Figure 4.3
Sister Elaine Roulet in 1991 at the Children's Center at the Bedford Hills Correctional Facility in Westchester County, New York. Joyce Dopkeen/*The New York Times.*

officers. Bedford Hills is now considered a "go home" prison, meaning it emphasizes rehabilitation programming that is useful in parole hearings—from education to parenting programs to addiction treatment. It is also a maximum-security prison. Being eligible for parole and seen as a reformable prisoner is therefore its own work once inside these institutions; even when these programs are in place, they are not necessarily accessible. Beyond "go home" prisons, these programs are by no means common: in 2021, only a dozen states had these programs in one or more prisons. Generally, it is understood that within these co-habitating units, outcomes are better for everyone: for the infant, because they might securely attach to their mother; for the mother, under the logics of carceral capitalism, because staying connected with the child reduces recidivism.[89]

Beyond Bedford Hills, mother-infant programs rely heavily on court referrals—and many courts are "unaware" that the programs exist. The application process is arduous, and women have sued (successfully) to be pre-admitted when pregnant (for decades, women have had to wait until giving birth, then complete the application process, making them ineligible). Further, applicant files could be "lost," or applicants could be deemed unfit by psychiatrists for a whole host of reasons. The services are therefore underused even though prisons are at over-capacity. Once in, children in many of them lack medical care: infants have almost died from delayed routine checkups in their first weeks of life. And many mothers complain that part of these programs is intense programming or reprogramming, that they keep their child only on the condition that they work and take endless classes; in essence they are partially kept from their child via mandated education even if the child lives with them. Mothers in the programs are subject to intensive surveillance and its resulting interventions; their children can be removed from the programs for any infraction or if the child's pediatric screenings result in any red flags. Those who gained admissions to the programs are overwhelmingly white, even though two thirds of incarcerated parents are nonwhite;[90] the notion that a good mother in prison is a white one has a deep and long tradition central to the naturalization of attachment science at mid-century and beyond.

As new scientific understandings of motherhood continue to be substantiated, parents in the carceral context remain central to their codification. Attachment norms are actively being researched in the present, and incarcerated mothers are still the subjects of experimental psychology. Much like the flexible uses of attachment theory in the courtroom, research continues both to defend the child from a possible future that is traced etiologically to a criminal mother (and her environment, the prison) as well as to argue that the mother is the total, essential environment of love. As just one example, Mary Byrne, a professor at the Columbia University School of Nursing, makes periodic visits to Bedford Hills, continuing the tradition started by Spitz in that same prison laboratory/nursery.[91] Once there, she and her team have used the "Strange Situation" procedure to assess the attachment types of the infants residing in the Bedford Hills nursery. Whereas historian Katie Joice argues that Ainsworth's work has lost discursive relevance for child-rearing,[92] we can see here that it still directly informs the right to mother in the penal context: the "longitudinal study had several aims . . . related to attachment, maternal qualities, mother-child interaction, child development over time, and maternal criminal recidivism."[93] Microanalysis and attachment remain central tools managerially deployed to govern how the incarcerated parent is allowed to bond; findings maintain that children within the program have rates of secure attachment (and disorganized infancy) comparable to rates in the community, taken to be the sign of a good prison nursery program.[94] Attachment theory is still used to substantiate the notion that all kinds of mothers might live with their children without doing them undue harm.[95]

PARENTAL CONTROL

By restoring this experimental and actuarial context to Spitz's work and its afterlives, we can see predictive theories of maternal bonding as a crucial hinge in the management of incarcerated mothers, where its norms of maternal bonding are elaborated by cleaving while also producing it—in both senses of the word. The regulation of mother-child interaction, where mother is understood to be both the only possible good medium and herself negatively impactful, originates from this range of mid-century theories as they met logics of recidivism and incarceration. As James's rich concept of the "Captive Maternal" allows us to understand, this was an extension of the chattel principle, reprised in the deployment of a punitive science of maternal fitness in the Progressive era, with its eugenicist panics about race and race futures (eugenics itself actuarial and predictive). Attachment theories have been part of this long project since the 1940s, with the development of a science of predicting juvenile delinquency from maternal emotionality and reading back from childhood behavior to maternal capacity. In our present, it continues to inflect this project as attachment science is rehearnessed to argue that women in prison do or do not deserve to mother—and that infants deserve to be mothered by their mothers or that they deserve the stability of a maternal environment in the mother's carceral absence.

The laws that support and curtail maternal bonding, and the logics that inform them, doubly revise theories of maternal care. Traditionally, we date the beginning of these predictive maternal theories and their elaboration to the 1940s and then see *them* subsequently ratified and applied to carceral settings via law some fifty years later. Instead, Spitz's work proves to be an evidentiary regime concerning mothers and children separated in prison. Because this mother-baby unit is under observation and control, "Captive Maternals" have been sorted, documented, experimented on. Given that prison nurseries and infant programs significantly predate the Reagan era, their resurgence suggests a form of resistance by incarcerated mothers, where they take a theory of fitness and stimulus predicated on their experiences and use it to maintain connection with their children. Taken together, theories of bonding and baby watching have always been about control a priori, if also susceptible to being leveraged by the very mothers both science and the state have sought to capture.

"A prison is defined by its rules of confinement."[96] It isn't just the fact of confinement that matters, but its nature: what materials, what thickness of walls, how congruent living situations are managed and surveilled, what types of sensory deprivation are in use, what kinds of isolation, what kinds of attachment are permitted, as well as why it behooves carceral capitalism sometimes to permit them. At the prison nursery of Bedford Hills (and beyond), the fundamental unit of correction, and the fundamental unit of predictive developmental science, is the mother-baby.

III THE MOTHERING OF MEDIA

5 A HISTORY OF SCREEN PARKING

In the summer of 1968, Mildred Colman, a member of the community council of the Crown Heights section of Brooklyn, called New York's WNEW-TV (channel 5) to issue a complaint: "there were little children running around at night getting into mischief... it was the fault of the parents."[1] "Mischief" is a loaded term, especially in the context of Crown Heights and "the long hot summer." First, Crown Heights was undergoing extreme white flight between 1960 and 1970, going from a population that was 70 percent white to one that was 70 percent Black.[2] Second, especially in the context of ongoing riots across the nation, mischief is a both an extralegal accusation and a criminal definition—a term of art for property destruction.[3] Colman did not call the police on the mischief-makers, or at least not this night. Instead, she called WNEW-TV's nightly Focus segment, and asked them to do a piece urging parents to "keep track of their kids."[4] This was ever more urgent, from her perspective, because of the curfews put into place across the nation during the riotous summers of the late 1960s—children should not be out of doors at all in the evening.

In response, Charlotte Morris, the station's public affairs director, went out to Brooklyn to make a TV spot featuring a Latina, a Black, and a white mother. The three women were lined up next to each other. The first mother, a parent of five children, said that she always knew where her children were—and then asked the camera, "Do *you* know where *your* children are?" The next two parents echoed the statement, "Do *you* know where *your* children are?"[5] It was an instant hit—the watching public wrote fan letters again and again—and the insinuation moved to the top of the hour: "It's 10 pm. Do you know where your children are?"

The public service announcement used television as a prescriptive technology, leveraging the old technology of peer shame to startle neglectful parents into doing their parenting, while conflating the work of parenting with effective surveillance. The phrase, in contemporary parlance, went viral and was soon featured on many local broadcasts around the nation. Mel Epstein, the director of promotions for WNEW-TV, approached the spot differently, retaining the early New York spin on the ominous question. He decided to have celebrities come into the studio, some as unlikely as Andy Warhol, who seemed slightly confused as he read out the famous slogan. In total, some 800 celebrities would record the message across fifteen years, and, on October 21, 1985, WNEW-TV hosted a party at Studio 54 to tape a live spot with as many of the alumni as they could gather, chanting in unison: "It's 10 pm. Do you know where your children are?"[6] Behind the cachet of celebrity, the spot had become a comforting ritual: the news opened this way across the 1960s and into the 1980s, a fixture, no matter the actual meaning of the slogan, or what news followed.

At the end of a day of broadcast, parents were reminded of their children, playing out in the street, yet it is another paradigmatic scene that, in this same period, transformed television's meanings in the home. If, in the evening, it broadcast a maternal message to other mothers, in the mornings, it was a blank slate, a test pattern, snow, static. Children reportedly, in perhaps apocryphal anecdotes, sat captive to the medium without a message. If, in the evening, the child was assumed out of doors, in the morning, they were largely already assumed to be in front of the television. The problem was, the television was not exactly "on"—it offered up nothing but a little set of diagrams, perhaps a stereotypical picture of an Indigenous man with a war bonnet. This was a test pattern, upon which the television could train and be adjusted. Now this static image took up the ritual space in the middle of the living room. Nonetheless, parents worried over the children passively looking at a collection of circles. This worry joined with that centered on those children out and misplaced at night. As it turned out, children were also misplaced in the home precisely because their attention was focused on the television.

Now, our media panics center on the tablet—usually the iPad. We call this "screen time," a term, as we saw in the introduction, coined in the 1990s. Not only are parents worried about screen time for children, but the anxiety about screen time, as Phillip Maciak argues, has spread from children to all users of screens. This indicates that we need to better understand the origin of child-and-media panics to understand the everyday media theory of users as they self-manage and regulate screen time. Put simply, the children featured in this book have grown up to be the parents who have worried over screen time since the 1990s to the present day. They are still worried about their own screen usage, and now also that of their children.

Screen time for kids often centers on a cost-benefit analysis: children are occupied and parents relieved—but at what cost? Pacifying children is important to domestic economies: it allows for other laborious tasks to be completed, but how do media (and *the* media) impact children? These have become central social and psychological questions. As we have seen, Americans debated the appropriate mode of parenting in an age of rapid technological change across the twentieth century, but *pacification* via media, as a way of obviating frustration in everyday life, has remained the most conflicted site of media-as-mother. Before we can have screen media in childrearing as a social problem, parents had to deploy this technique of absorptive care. Using screens for pacification and occupation of children—what I call "screen parking"—has a long genealogy. Often thought of as emerging with television, media panics centered on children's absorption have existed at least since the optical toy, and have been attached to radio, various print media, and the cinema.

Worrying about children looking at test patterns—indulging in the medium—might seem quaint. But the very conditions for contemporary parenting concerns over hand-held, individualizing media, consumed by one child on one small device, germinated in mid-century concerns over television. As we shall see in this chapter, if in the 1950s television was a medium rife with contradictory social meanings, across the 1960s, the television became understood as a potential vehicle for better, more attuned programming. Television became largely synonymous with the fantasy of achievable equality via distanced care and education. As the scene of domestic care changed for many middle-class families in this period, by the 1980s, television was understood as a proxy-mother, a proxy-educator, the "boob tube," the electronic babysitter. The condemnation of mothers who used media as proxies was widespread, and yet retained specific classed, gendered, and raced taxonomies. Mothers who worked or were divorced, as we will see, were often invoked as the most likely to *over-rely* on screen parking, and as media theorist Tung-Hui Hu shows, the rise of the "couch potato" is highly racialized and co-emerges with the trope of the "welfare queen." As I will argue in this chapter, the intervening ten years or so were a crucial decade for the meanings and uses of television. This chapter asks how various media were understood as impacting children, and how our screen media, particularly television, became media-as-mother. Throughout, from the early Payne Fund Studies on emotion, influence, and cinema to the mid-century battle between psychiatrists over comic books to the golden age of children's TV, we will see how the child was reconceived of as a media effect, an outcome of proper or improper stimuli. The fantasy that a television set, and, following Lynn Spigel, the ritual space in front of it, might standardize child development was held up as a tantalizing ideal that was also castigated from all sides.

As media were a set of contested tools for pacifying children, child development specialists argued about how television and other media affected the American child while in their absorptive, if not catatonic, state. In previous chapters, I argue that the metaphorization structuring bad mother theories at mid-century allows us to think of psychologists as offering as a media theory of mothering. This implicit theory rests on an understanding of maternal qualities—her attention, her presence, her environment—as constituting a medium of care better thought of in terms proper to media rather than sentimental mothering. This chapter works from the other side, to elucidate how media (and the content of same) became understood as a central concern of mothering. To the earlier chapters I add here that media too were increasingly read as having the same qualities as a bad mother, and that these media mothers became another site of intense scrutiny. If mother-as-medium is a crucial paradigm for parenting across the twentieth century into our present, it might be no surprise that panics surrounding media might also center on the possibility of media-as-mother. As we shall see, cinema, comics, and television were each attended by a social media panic that centered on children (although all three mediums were co-present by the 1940s and formed "a media ecology"). Cresting when television became the dominant medium of speculation and fear, media researchers began to ask what kinds of subjects are made by media and when. They were joined by pediatricians and psychologists in a choral question not that different from WNEW-TV's nightly query: Not where, but who, is television's child?

MOVIE-MADE AMERICA

By the 1960s, when social scientists and psychologists turned to the problem of children tended by media, they were able to draw on four decades of research on the impacts of media on emotional life. Since its earliest professionalization, media researchers, starting in the 1920s and 1930s, were as much studying media—particularly cinema and the radio—as studying their users. While the lion's share of early media research about the radio focused on mass media's *mass* reception, especially as propaganda, researchers looking at entertainment—particularly cinema and radio—often took children as their object of focus. In their vulnerability, children were understood both as in need of protection from the assuredly harmful impacts of new media and as showing more baldly the effects that those media had on adults. But propaganda and entertainment were not yet distinct categories. Both looked at how media were taken up in the mind and how they influenced behavior.

Indeed, mass suggestion was understood as the signal media effect in the 1920s, called the "hypodermic" effect, wherein media pierced the individual mind (which, as debates subsequently tell us, may not even have been believed at the time and may have been a straw man). Radio might be considered a father-medium—authoritative, reliable, communicating directly and intimately in the home or, as McLuhan would have it later, a medium that allowed for "an ancient experience of kinship webs of opaque tribal involvement." As by now should be apparent, cinema was multiply charged otherwise.[7] Movie-going took children beyond the home, into the dark, and into the masses. Children then had experiences beyond parental control with a new medium, which was presumed to hold enormous power over child and adolescent development; the movies might remake children in their image.

As the 1920s progressed, media panics met the energies and attitudes of the Progressive era, and a turn toward the effect of media on children became urgent. As Garth S. Jowett, Ian C. Jarvie, and Kathryn H. Fuller show, the Progressive era's many actors were increasingly turning to science to settle moral questions.[8] There were therefore many stakeholders in determining how a child—and how an adult—might react to media and its undue influence: the clergy, educators, pediatricians, psychologists, sociologists, and, yes, parents themselves. Charged with determining the impact of media on children, the early media researchers (largely sociologists and psychologists) were now aided by new tools for measuring shifts in emotionality. The largely used psychogalvanometer had been developed by psychologist John A. Larson for the police in Berkeley, California. From its roots in the juridical, it offered a sheen of evidentiary truth—the body telling on the subject's mind. By 1921, the tool measured heart rate, breathing patterns, and blood pressure across time.[9] The device helped batch-process subjects, but, as historian of technology Brenton J. Malin shows, its introduction protected psychologists and sociologists alike from the charge of subjectivity that attached to much of their research in this period.[10] Scienticity was to be assured via a turn to observable phenomena, statistics, and the verifiable measurement of sensation.[11] Nonetheless, subjectivity reappeared in morality.

Between 1929 and 1932, morality and scientific inquiry found their way to one another in what would become known as the Payne Fund Studies. The Reverend William H. Short, a key figure in the founding of the Payne Fund's Motion Picture Research Council (which would go on to introduce a series of self-censoring and eventually binding codes for film production), sought empirical evidence of cinema's moral rot and its negative impacts on youth. He approached Robert Park, one of the most prominent sociologists at the University of Chicago (which had the most prestigious sociology department in this era, especially as concerned "urban research").[12] Park in turn recruited a number of his own students, and the Payne Fund Studies began, taking as their hypothesis that cinema fostered degenerative social morality—despite the fact that its sociologists already had supposedly determined movie-going could be implicated in only 10 percent of criminal activity (which, it cannot be overstated, would make it a *huge* driver in criminality).[13] Payne Fund

Figure 5.1
Child pictured with psychogalvanometer on the table and a Kodascope on the stand at the back of the room. From Wendell S. Dysinger and Christian A. Ruckmick, *The Emotional Responses of Children to the Motion Picture Situation* (New York: Macmillan, 1933).

Studies—located in college towns like State College, Pennsylvania; Cleveland, Ohio; and New Haven, Connecticut—produced twelve individual volumes on "Film and Youth," which were collected and published in 1933. Each took an explicit focus—from how children slept after seeing movies (conducted overnight in a school, with subjects hooked up to psychogalvanometers) to how children learned criminal behavior from films. Figure 5.1 shows a psychogalvanometer in use.

While some studies aimed to project the nation's movie attendance (from the Ohio studies, say), others determined that children who went to the movies had disturbed sleep, and measured these with jiggleometers—a tool for tracking fidgeting and small motions—attached to boys who slept barracks-style overnight, in a high school turned sleep laboratory. Most of these studies concluded that, in short, movies caused an increase in delinquency. The "movie problem" that begat the research was substantiated with its findings.

This should not be surprising. Reverend Short approached Park for exactly this reason; urban sociological research focused on "delinquency" in this moment. The studies were set up to control *moral* contagion—media understood as interfering with the good upbringing of children. All media were, in essence, a bad mother on the Payne Fund Studies' understanding, one who taught the wrong lessons at scale to auditoriums of children nationwide. That movies could *teach* was the proof of their contagion. But how they taught, rather than what they taught, and how that educative function might be harnessed, was more of an open question than the settled science presented by the twelve studies. The Payne Fund Studies were not only interested in movies as content. Researchers at various sites wanted to gauge dosaging of movies (did seeing them more or less change criminal impacts?) but also, presaging the test patterns that concerned parents some thirty years later, showed children the mere flicker of film to see what the *medium* was doing to children absent of image or narrative.

A suppressed thirteenth study shows, in its contrast to the published twelve, that there were other scientific modes of reading media and its effects.[14] This case, which considered children living in poverty in East Harlem, pseudonymized as "Intervale," was helmed by Paul Cressey, an MA student of Park who had completed a PhD at New York University and moved on to a professorship at Evansville College. Cressey was a late-breaking addition to the Payne Fund team, joining toward the project's conclusion. His work undermined the rest: it argued that films offered an "informal education" and were not a site of direct influence. Moreover, Cressey thought that public schools did not necessarily care for or cater to their constituents—they were, in short, sites of alienation and fragmentation. Whereas Park and his students thought of school as the bedrock of civic life, and media as a source of corruption, Cressey largely reversed this understanding. If the general studies pitted schooling—that disciplining force for good—against Hollywood and its impressionable wayward viewer, Cressey thought that schools could isolate or scale marginalization while cinema (or "photoplays") could as easily mold children into the right kind of "citizen."[15]

As media historian William J. Buxton details, this was not only because the study occurred in New York and among impoverished participants, but because Cressey was able to escape the "unchallenged" views of Park and his two devoted students, Blumer and Hauser, carrying out Park's indefatigable orthodoxy.[16] Operating at these distances, both material and intellectual,

Cressey ventured that what a "photoplay might mean" was accessible only through the psycho-economic and social details of *each* individual who might see it in a darkened auditorium. Rather than looking for what would become known as media effects, Cressey was decades ahead of his peers and his teacher in his understanding of media. Cressey was, in effect, disputing causality—his study was suppressed. The published studies ostensibly took media as the site of effect; media were the explicit subject, and controlling and patrolling media was the studies' primary aim. The children simply served as a register for cinema's power. While the lions' share of the Payne Fund Studies advanced the direct-suggestion view, what I want to suggest here is that a conflation of subject and object was already underway in the design of the studies.

The Payne Fund Studies are of interest not only because they inaugurate the tradition of psycho-sociological media research but because they show that media effects research has long taken interest in children as an index of media's corrosive effects. This research braided ongoing psychological and sociological theories of childhood pathology with new emergent research on propaganda and mass media. Mass media, then, were, in the wake of the Payne Fund Studies, understood to offer uniform impacts on citizenry; who might be a citizen and what the effects might be were flattened and universalized. This was the "magic bullet" theory—soon to be disproven when the human subjects were shown to be active interpreters of their own media experience by Hadley Cantril, Elihu Katz, and Paul Lazarsfeld.

Children have, by and large, been treated as though they mimetically translate life on screen to their own persons (even as those screens and what they offer have changed greatly). There is a kind of implicit script theory, one that moved with media (or was not medium-specific). Where the Payne Studies do not particularly look at the familial milieu in their sociological account of media-child relations in the 1930s, but instead looked at cinema's interloping in the family, by the 1940s, the focus on the mother-as-medium had begun to emerge. The constellation of media and maternal effects began to render them indistinct. During the domestic containment years of the early Cold War, media psychology and childhood psychology would knit together further. As the movie-going panic settled (and the Hays Code, as a note, was operant from 1934 to 1968, keeping movies supposedly in check), a new medium appeared upon which psychologists could affix what the Payne Studies called a "legacy of fear"—that media were a source of social learning and therefore social contagion.

SEDUCTION THEORY

The Payne Fund Studies gave empirical credence to the idea of cinema as a corrosive force (while suppressing a more complex view), but other media began to attract substantial attention from psychiatrists across the 1940s and 1950s, who moved from the notion of media as contagion to one of direct, operant conditioning (see chapter 2). At the center of the panic: comic books.

If mothers indeed impress themselves on children, it is because children are understood as impressionable. The moral panic about the medium of comic books—which stood in opposition to the morality of the postwar nuclear family—indeed centered on that impressionability of children. As with the Payne Fund Studies, the comic book panic started out in religious quarters before gaining steam in the mainstream press. Even in 1940, editorials were published that accused the medium of being bad for young bodies and minds. Sterling North wrote in the *Chicago Daily News* in 1940 that comic books were "a national disgrace," demanding regulation of the form. He disparaged the medium wholesale: "Badly drawn, badly written and badly printed—a strain on young eyes and young nervous systems. . . . Their crude blacks and reds spoil the child's natural sense of color; their hypodermic injection of sex and murder make the child impatient with better, though quieter, stories."[17] For North, comics as a medium ruined children's bodies via their physical production ("badly written and badly printed"), and their message turned them into delinquents. His stance expresses the same dual focus on physiology and psychology seen in the Payne Fund Studies. Now, in the postwar era, conditioning was understood as opening up children, at scale, to delinquency, to communism, to queerness, and to other outcomes deeply pathologized by the conservative human sciences. The familial medium was not enough to secure children for democracy—mothers were failing at scale. Recalling the hot and cool mother paradigm, comic books were here misunderstood as a hot medium (McLuhan would call them a *cold* medium for their underdeveloped nature): they did not spark the imagination of the child. Instead, they flooded the child, wore down their senses, and then, with the child undefended, led them to literal belief in graphic depictions. Or so the argument of the panic went.

Just as the Payne Fund Studies were brought in to scientize the Motion Picture Research Council's conservative anti-media panic, the comic book panic solidified in the late 1940s only when a scientist gave extraordinary credence to such fears in public. In a 1948 story entitled "Horror in the Nursery" in *Collier's Magazine*, and another in *Ladies' Home Journal* in 1953, psychoanalyst and psychiatrist Fredric Wertham offered a scathing take on the comic book—one that rhymed with new pathologies of motherhood circulating in the postwar period (see chapter 3). This was no accident.

Wertham had been born in Nuremberg and worked at the Kraepelin Clinic in Germany, where Adolf Meyer, who would be an early mentor once he emigrated, had also trained. Wertham was long invested in a wide range of political and literary texts beyond Freud—as political theorist Kevin Duong puts it, "Marx agitated his head, Dickens his heart."[18] After becoming one of Meyer's many protégés at Johns Hopkins, Wertham moved to New York in 1932. Duong again, "By then only thirty-seven years old—he was already caught up in three world-historic currents at once: Marxism, psychoanalysis, and the agitation for democratic equality for Black Americans."[19] Wertham was also moving across networks that placed him immediately in dialogue with many of those same, mainstream actors—including Mayer—who were developing the idioms that constitute the mother-as-medium paradigm.

With one foot in normative, prescriptive theories of mind and mother, and the other in an alternate and radical lineage, Wertham had a robust career as a leading New York psychiatrist. He eventually ran Bellevue's Department of Hospitals program before becoming the founder of the radical Lafargue Clinic in Harlem in 1946. Posed directly as an alternative to Bellevue's own racist science of mind, the Lafargue Clinic would serve the impoverished Black community that surrounded St. Philip's Church where Wertham saw patients. The clinic gained repute with the backing of Wertham's former patient Ralph Ellison as well as novelist Richard Wright (as Wertham wrote: "The Freudians talk about the Id / and bury it below. / But Richard Wright took off the lid / And let us see the Woe").[20]

Wertham's politics were rare for white psychologists interested in childhood development in this period: he was interested not only in Freud's medically relevant family as the site of neurosis but in the larger social view. Social inequality, and particularly racism, could and did suffice as an etiological explanation for neurotic behavior. Wertham and his team, including Hilde Mosse and Ernst Jolowicz, were keenly aware of the limits of making a psychoanalytic clinic in an impoverished neighborhood. Like the adaptations that the second generation of psychoanalysts in Vienna and Berlin had used to treat widespread war neurosis in the interwar period, these clinicians transformed analytic techniques into those that might batch-process patients—namely, turning to group play, open case presentations, lectures, and other methods.[21] As Kevin Duong describes it, "The technical compromises imposed on the Lafargue Clinic's daily work were so far-reaching, and so numerous, that at a certain point it simply stopped looking like psychoanalysis."[22]

Wertham was rarer still given his prominence in the American public eye. NBC went so far as to pilot a script for a teleplay, "The House I Enter: The Story of an American Doctor," which seems to never have been produced (and once Wertham took aim at comic books, he was lampooned regularly in them).[23] But before he would take aim at media, Wertham took his experience and theories centered on the psychosocial consequences of racialized poverty, and testified as a key expert witness in NAACP's fight to end school segregation in *Brown v. Board of Education*. Yet despite his work on the psychosocial—which has recently drawn attention once again—Wertham is best remembered for instigating and reifying the comic book moral panic of the 1940s and 1950s.[24]

If Wertham's work in the Lafargue Clinic was radical in its moment—and keenly interested in the neuropsychological impacts of poverty—his comic book commentary in the public was highly conservative. These apparently incompatible stances were paradoxically dependent on one another: it was the children of Lafargue Clinic who stood as his first and best evidence

Figure 5.2
Fredric Wertham reading a comic book in his office. Photograph by E. B. Boatner, Schomburg Center for Research in Black Culture, New York Public Library.

in his work on comics. He wrote frequently in the press, from *Ladies' Home Journal* to *Collier's Magazine*, about the ills of comic books and what they were doing to American children. Often, he used data collected at the Lafargue Clinic as evidence (more about this below). While comic book burnings spread rapidly following Wertham's *Collier's Magazine* piece, the comic book panic reached its apotheosis when Wertham turned from his earlier work on psychosocial causes of delinquency (poverty and racism) to a new theory of mind, which argued that comic book violence was instigating the same (see figure 5.2). Building on his writings for the public, in 1953 Wertham released *Seduction of the Innocent*, which was read widely and seriously by American parents. The book in part led to Senate hearings on comic books and a new comic book self-censorship code—much like the 1930s film industry adoption of the Hays Code.[25]

Seduction of the Innocent is a psychosocial investigation into the increase in delinquency that opens with a parable drawn from Wertham's clinical experience. In the early summer of 1950 in New York City, there was a shooting at a polo grounds. After placing a dragnet across the city, a defendant was found, a Black teenager called Willie. At Willie's apartment, several guns were recovered—and Wertham details his confession to the police that "he had owned and fired a .45 caliber pistol—which incidentally was never found."[26] Wertham tells us that Willie was then sentenced in court, where a judge stated, "We cannot find you guilty, but I believe you to be guilty." Willie was, at the time of Wertham's 1953 book, serving an "indeterminant sentence in the state reformatory." Another, similar shooting occurred shortly thereafter. Although Wertham never directly says it, it is clear that he believes Willie innocent of this crime, which makes it all the more strange when Wertham turns around and tells his audience two things: first, that Willie was in fact Wertham's patient in Harlem at the Lafargue Clinic since before Willie's second year of life, and thus Wertham knew every detail of his development, and second, that he will take Willie's story as a case study, as concrete psychological evidence—despite being innocent—for a mass psychogenic delinquency due to comic books.

The Lafargue Clinic was a key staging ground for gathering evidence of the widespread influence of comic books on children, priming them for violence: case after case was brought out, with each child subject to a battery of tests—Rorschach, Duess, and Implicit Association tests most prominently. Wertham was concerned with increases in anti-Black and antisemitic racialized violence due to depictions of that violence in comic books, and worried about the conditioning of comic books *toward* fascism (he read *Superman* as an extension of Nietzsche and argued that it also was convincing children to be homosexual). But there was also a running current in Wertham's book: the problem of delinquency.

As we saw in previous chapters, the 1930s and 1940s had been rife with emergent theories centered on delinquency in young adults. Wertham himself described the literature thusly: "One might think that society hopes to exorcise [delinquency] by the magic of printer's ink. . . . Juvenile Delinquency holds a mirror up to society and society does not like the picture there."[27] Wertham was *against* recrimination of mothers and children, and the kinds of bad mother pathologies appearing with alarming frequency in the popular press and in psychology. Without naming it, Wertham most likely is speaking negatively about John Bowlby's psychological work on the delinquent published in 1944, with its titular study of forty-four juveniles who, between the years of 1936 and 1939, were treated at the London Child Guidance Clinic in relation to their propensity to steal (see previous chapter).[28] As Bowlby advanced what would become his theory of attachment and its predictive nature for child outcomes, his juveniles (some as young as five years old) are shown to have suffered separations, institutionalization (in foster care if not clinical settings), and often bad treatment along the way. Wertham put his shoulder to the gears of this mainstream theory, the neat formula that the delinquent child, the bad child, expressed a broken maternal bond. He did not think that the wrong kind of mother produced the wrong kind of child, but that the wrong kind of society, and its media, did.[29]

Therefore, how Wertham approaches comic books is puzzling, at least at first blush. Wertham approached the question at the core of *Seduction of the Innocent* with one difference from Bowlby: if the bad mother is not the root cause of mass child delinquency, with the numbers of children in reformatory sky-rocketing, might it be bad media? First, Wertham retains his psychosocial theories. He nods to the fact that judges are overly happy to send children to reformatories—which Wertham saw firsthand in Bellevue, and argued that they do so with "a light heart and a heavy calendar." But then Wertham makes his pivot, and tells his reader in no uncertain terms the responsibility lies with comic books, which educate children who then, when they face overly harsh sentences, go "onto a long postgraduate course in jails (with the same reading material)."[30]

Wertham argues that it is because of childhood "unprotectedness" that they become criminal. Yes, children are exposed to violence in the family (which he mentions only in passing), but, he argues, it is only when exposed to violence in media that children become mimetically aggressive. Wertham advances an argument in which media have supplanted *his social theory* as the site of pathological etiology.[31] He slips his comic book media into where Bowlby's mother should be. The end result is the same. For a radical psychiatrist, deeply aware of the effects of racism and poverty on mental welfare, and advancing a psychosocial theory of delinquency (while indeed concerned over increased racial violence as part of delinquency) when few non-Black psychiatrists were interested beyond eugenics, this move is a contradictory one. Wertham turned away from the family as the social and emotional driver of health to something systemic, but rather than choosing the social forces that he saw present in his clinic, he made the comic book the primary vector of social contagion. Wertham went for an appealing panic, one centered on media, as a way of avoiding going for the mother panic that has consumed his colleagues and their research. In doing so, he ended up subbing out one environmental panic for another, taking on the terms of the mother panic, but attaching them to media.

Wertham was then in the grips of a panic partially of his own making—and it had far-reaching effects on the comic book industry, and inflamed parents. But its evidentiary basis—which was part of why Wertham had such great reach—was one thoroughly debunked by later scholars. Media researcher Carol Tilley even compared Wertham's patient records from the Lafargue Clinic with the case studies in the book, only to find Wertham exaggerated the entire basis of his case: the number of patients he saw, their statements, recollections, dreams, test results, ages, and biographical details.[32] He had taken the data of his psychosocial project, and then transferred, scaled, and amplified it to make a media panic that would consume parents.

Wertham was not alone in looking at comic books as enemy number one. There was a veritable cottage industry of expert psychologists serving as witnesses and consultants both for comic book producers and in association with social groups standing against comic books—all of whom thought it a powerful medium, whether for good or ill. Dr. Jean A. Thompson, head of the Bureau of Child Guidance in New York City; Sidonie Gruenberg, director of the Child Study Association of America; and Dr. Lauretta Bender, head of the children's ward at Bellevue Hospital, all staunchly defended the good to which a conditioning medium can be turned.

Bender provided a diametric opposite to Wertham's more well-known and negative theory of media-as-mother. She too derived her sense of comic books from her clinical work at Bellevue Hospital. Bender was an MD by training, and an anti-Freudian (despite going on to marry a leading New York psychoanalyst). Bender had continued her medical training at

Johns Hopkins, in the same period as Leo Kanner's emergent work on autism and overlapping with Wertham—studying ninety schizophrenic women in the same clinic. Bender was the first to argue against the notion that a mother—or any family system disorder—caused autism, insisting instead that it was neurologically based. When Bender left Johns Hopkins for New York City, she took up a position at Bellevue, and by 1934 had ascended the ranks there to senior psychiatrist in the children's ward (in 1936, the first public school for Bellevue's children's population opened within the hospital). There, Bender worked with 250 children, whom she classified as having two forms of neurological difference: those with "Heller's Disease, because they seemed to develop normally to three or three and one-half years of age and then regressed without any known reason, such as an illness, or any neurological signs. The other group did not develop beyond the two-year level."[33] Bender presented these findings in 1940 at the American Psychiatric Association, and soon thereafter, she recalled, Kanner wrote from Johns Hopkins to ask her, "How did [she] ever find such children?"[34]

The answer was, of course, that they were in the wards of a public hospital. While Kanner would go on to study exclusively white children at Johns Hopkins (see chapter 3), and then link the racial and class distinctions of these children's mothers as highly related to their resulting diagnosis, Bender drew on her work at Bellevue, which, like at Lafargue, was largely with Black children in the 1930s and 1940s.[35] Indeed, Wertham had set up Lafargue, in part, to *prevent* children from ending up in Bellevue, where they were treated by Bender. This included the 250 children that Bender saw in the clinic, who were considered "emotionally disturbed" prior to further refinement of childhood psychiatric diagnostic categories, which she presumed were all neurological in etiology. Bender thus used chemical cures—from insulin overdose and short-term induced coma to electroshock therapy—on the children she cared for (she has now been greatly criticized for these methods). By Kanner's 1943 publication, which joined with other family-centered etiological approaches to autism, Bender's non-maternal theory became a minor tendency. For each of these lead psychiatrists, Bender and Wertham, a refusal to pathologize the Black family via the theories and inheritances of Freud involved a crucial shift: for the former, one had to turn from the mind to the brain, and for the latter, from the family to the level of the social.

Along the way, Bender, like Wertham, also studied the impact of media on her subjects even if this was not her initial interest in their treatment, writing and presenting several findings that ran contrary to Wertham's.[36] After *Seduction of the Innocent* reified the comic book panic, becoming its locus of authorization, Bender and Wertham faced off in 1954, when televised Senate hearings took place on comic book regulation before the Senate Subcommittee to Investigate Juvenile Delinquency in New York City, in what would be known as "the Kefauver hearings." Bender, who was quick to note that she had "quite a number of associations," listing her medical credentials (although by this time Bender was also on the editorial board of National [now DC] Comics), promised the committee that she had long been interested in the good powers of the comic medium before she had "any connection whatever with the comic people."[37] In the testimony that proceeds from clarifying the relationship of her fiduciary partnership to her clinical findings, Bender offers that it was via her young, "emotionally disturbed" patients at Bellevue, "when we were hard put to find techniques for exploring the child's emotional life, his mind, his ways of reacting, when the child was separated from the home and brought to us in the wards at Bellevue, [that] I found the comics early one of the most valuable means of carrying on such examinations, and that was the beginning of my interest in the comic books."[38] Although she argues positively for the clinical value of comic books (akin to a kind of play therapy), Bender got onto shaky ground with her clinical examples when asked about her role at National Comics. The chairman asked her what Superman might want with her, and she replied that her role was to think about psychological impacts of the comic books, such as the mimetic desire for flight in children (the licensing company initially refused to sell Superman-branded capes in stores, as they feared children would hurt themselves trying to fly but consulted with Bender nonetheless). She agreed avoiding a lawsuit was wise: "[W]e frequently have on our ward at Bellevue the problem of making Superman capes in occupational therapy and then the children wearing them and fighting over them and one thing or another—and only about 3 months ago we had such, what we call epidemic, and a number of children were hurt because they tried to fly off the top of radiators or off the top of bookcases or what not and got bumps."[39] The committee, having gotten what they wanted out of her—a testimony in favor of the danger of comic books—excused her and thanked her for her time.

When it was Wertham's turn, he argued the opposite, as he does in his book, similarly using empirical evidence of the impacts of comic books from the clinic. He then added, "All this to my mind has an effect, but it has a further effect and that was very well expressed by one of my research associates who was a teacher and studied the subject and she said, 'Formerly the child wanted to be like daddy or mommy. Now they skip you, they bypass you. They want to be like Superman, not like the hard working, prosaic father and mother.'" In Wertham's account of delinquency, children had exchanged mother for medium.

HOMICIDE AT HOME

At the end of *Seduction of the Innocent*, Wertham devotes a chapter to television, which he termed "homicide at home."[40] If comic books, like the movies, were ostensibly rotting children, one problem Wertham pointed to was parents who misunderstood the medium or its content, thinking erroneously that because comics were *books* they were innocent, just little illustrated stories about animals, when they were all blood, violence, and sex. This was in part because comic book reading took place apart from the family—in school and after—much like the cinema was beyond the home. Children in this period were increasingly understood as safe when home and unsafe when not. In the 1950s, as users of the television interacted with the medium, played with it, documented it, and increasingly, in the words of Lynn Spigel, "made room for TV," television took on two, contradictory meanings with regard to its impacts on family life. It was, as she shows, understood to "bring the family together," to give husbands and wives a new object of shared interest (Spigel even quotes women saying that the television increased their sex life via "necking"). Also important, as Spigel shows, televisions were used by mothers to corral their children, bringing them off the streets and into the living room, where mothers could hover nearby and know exactly where their children were (presaging the television spot with which this chapter opened).

But in the late 1950s, stories about these same scenes gone awry are also fixtures in women's magazines and parenting literature. What brings the family together in some cases is also termed in the strongest of language "addiction" to the "electric babysitter." One example that typifies this ambivalence toward television is a spread in *Redbook* magazine in 1957. On the right-hand side of the page is featured a recurring column, "You and Your Child," by Bank Street educator and children's author Irma Simonton Black. As was typical for Black's work in *Redbook* and elsewhere, she addressed herself directly to mothers, here, listing all the available children's programming, noting what is acceptable and what is harmful. This list encourages television watching, or at least some of it. But the list serves as an insert in the magazine's five-page spread on the child who watches too much TV. Again, parents are delivered a contradictory message: A mother *should* use the television—but the right programming. She should also not use it too much. The larger spread was replete with illustrations of three listless children parked in front of the television, entitled "How Much TV Should Children See?," by Robert Gorman. There he argued that the use of television, especially a great deal of it, is indicative of a kind of maternal ambivalence, writing, "Why does a mother, who lovingly guides her child in every other activity let him spend a third of his day *unsupervised* by the television?"[41] Gorman continued, "You don't have to be 'against' television to agree that this ever-handy pacifier, bribe, and babysitter is being shamefully exploited in millions of homes."[42] In creating a slippage between TV and ad hoc childcare, Gorman accuses the otherwise good enough mother of neglect. Only she—being ever attentive—will do. He then calls the parenting behavior as he sees it: it is tantamount to "abuse." The child is left both unsupervised and absorbed, penetrated by the violence he sees on TV, stationary and safe physically, but mentally in peril. The father wants the mother to actively mother whenever her child is home.

What was being negotiated in the pages of women's magazines—the uses and impacts of television on children—was also being researched by psychologists and psychiatrists. Despite Gorman stipulating that all child experts were against television, the Council on Mental Health of the American Medical Association had been, by 1957, researching the impacts of television on children for years—and could not discern its impact either way. As with comic books, some psychiatrists thought the medium positive, and others linked it to the ongoing panic over unprotected children. Returning to the Senate hearings in which Wertham and Bender faced off, both psychiatrists linked comics and television. Where Wertham saw horror, Bender saw mere entertainment and not a hypnotic medium—if of little value. As she said, "The children also will frequently tell me—for instance, on television, I have to listen to it with my own children occasionally and I am aghast, 'My God, how can you stand such things, children?' They say, 'Mom, don't you know it is only television, it is not real.'"[43]

But if children could stand watching it, Bender and many other child experts saw TV as an object of entertainment that the mother could control, that would keep children off the streets and safe. TV's absorptive qualities were not only seen, therefore, as a negative affordance. Just the same, others saw TV as following in a lineage of the "movie problem" and comics—simultaneously configured as an interloper *in* the home (more reminiscent of the nanny, there to supplement and pacify child for mother) and an educator in the world of violence.

That films could teach and radio might control were settled science when the television began to glow up American living rooms. But that television gathered the family, focalizing it, was seen as a great benefit for those raising children. As Lynn Spigel writes, "The ideal of family togetherness that television came to signify was, like all cultural fantasies, accompanied by repressed anxieties. . . . Even if television was often said to bring the family together in the home, popular media also expressed tensions about its role in domestic affairs."[44]

As we have seen, what is at stake in the Payne Fund Studies, in Wertham's exaggerated *Seduction of the Innocent*, and now in the television debate is the question of whether the depiction of fantastic things—the "not real"—can cause children to break free of the structures that make reality—the family, for instance. As Spigel shows, mothers can regulate exposure because they run the household from within; in her function, she became attached to the appliance upon which that regulation depended, the television, which permeated her domain.

For Wertham, television was essentially slightly less bad than the comic book. He argued that when engaging either medium, children had the "entertainment flow" over them.[45] With cinema and television, Wertham argued there was a key difference: other people. Enjoying entertainment with others meant it was, to borrow from Marshall McLuhan, whose work was associated with such debates, a *colder* medium (see chapter 3). Not actually cold, but less hot. The viewer in essence had to work harder—the medium got inside less easily, and the child was in a less absorptive state. Although Wertham thinks of the television as a pacifying device—much like optical media had been used with children at the turn of the century—there is an assumed mother somewhere nearby and she would have known what her child was watching at any given moment.[46] The television, if nothing else, is an open secret. Nonetheless, he concluded, "I have found that children from 3 to 4 have learned from television that killing, especially shooting, is one of the established procedures for coping with a problem."[47] The younger the child, the more their passive absorption of media troubled him, the more hot media seemed to run.

About one thing, Wertham was certainly correct: in this period, if there was a television on in the day, it was likely that mothers were watching, too. Television was, according to CBS, "Where the girls are." As Elana Levine and other feminist media historians show, across the 1950s and 1960s, mothers were assumed to be watching TV while working, using the television as episodic background noise—and television producers fit the pacing of shows to this rhythm of labor.[48]

If mother was watching television in this period, the panic about her displacement might be read another way: that television was having intense *effects* on children, effects that had previously been understood as maternal. If everyone was watching the TV, then, as we saw portrayed in *Redbook*, no one was watching the children or, worse, the TV was. Even before television saturation was total in 1960, the TV was described as ruining the child via both its physical properties and its content—as with the comic book just years prior. The television destroyed children via its "glow" and "heat." If we recall chapter 3, mothers in this same period were understood, metaphorically, to be doing the same. Children were left "bug-eyed" and weak, robbed of the ability to think for themselves and rendered unable to distinguish reality from fantasy.[49] As we saw in earlier chapters, this too was previously understood as a mother's job, to help children negotiate what was real and what was not, what was socially and morally right and what was not. Eventually, McLuhan too would go on to argue that television watching for children was overstimulating (in his parlance, a hot medium) and even "addictive."[50] The presence of others, he decided, did nothing to destroy the absorption of the TV set.

At the start of the 1950s, only one in ten Americans owned their own television. Even so, panic around the new medium set in. In 1951, anthropologist Earnest A. Hooton described TV as "a visual education in how to do wrong." By 1954, when Wertham published *Seduction of the Innocent*, television was increasingly domesticated, an assumed feature of daily life (this was the year the TV dinner was introduced). More and more of family life was spent around the electronic hearth. By 1955, nearly nine in ten families had at least one television, and teenagers were shown to be spending thirty hours per week watching its limited programming. By 1959, 85.5 percent of US households reported TV-owning, making the appliance slightly more common than even an indoor toilet. Once the appliance became as nearly widespread as the myth of the nuclear family itself, television became a central site for developmental panics along these psycho-physiological lines.

The scale of the mother problem gripping psychologists across this decade was seen as an effort to strengthen domestic containment in the face of the Cold War. It found its complement in the scale of the media question, which borrowed its phrasing from Cold War containment (foreign) policy. As we saw in the preceding chapter, mother-centered psychology similarly borrowed from media discourses of its time to produce the containing effect. Influenced still by propaganda research and theories of contagion, media researchers decrying the "rot" of the American child due to mass media were now suddenly as omnipresent as mother-focused panics. To underscore, according to psychologists operating in this moment, mother and media were understood in similar (and sometimes purposefully interchangeable) terms, with similar primacy, each a force that could make its impact nationally at the level of one to one. The family and the screen scaled up to the question of the nation.

MEDIA-AS-MOTHER

By the 1960s, on the grounds of television watching in the absence of mother, the media-as-mother panic had begun to take shape. The latchkey child, soon to be a cultural trope and part of Gen X's identity, was at its center. When Josh Harris was a little boy in the 1960s, he would be placed in front of the television and left to watch it, like many other children his age. Like other latchkey children, he went to school and came home to the TV.[51] His father was away working (for the CIA no less) and his mother was employed as a social worker. Harris says simply, "The TV was my mother."[52] It makes a certain kind of vulgar sense, then, that Harris grew up to be the founder of Jupiter Communications and Pseudo Programs. Pseudo, which had its notorious offices in SoHo in the early 1990s, pioneered live-streaming television over the net. There, Harris hired young artists to host their own streaming shows that allowed audience members to comment in real time on what they were seeing and allowed viewers to interact with one another over the platform.[53] Pseudo also offered Harris the auspices under which he authorized his own networked art experiments, starring himself. In one such experiment, Harris and his fiancée lived together under constant, live-streamed surveillance—for a year. Anticipating MTV's *The Real World* and similar mainstream reality TV programming on the verge of emergence, cameras were installed in every conceivable corner of the apartment (the shower, the toilet, the bed) and transmitted the couple's life via Pseudo, 24/7. It may come as no surprise that the couple broke up before the year was out and that his fiancée experienced this streaming as psychologically devastating. But perhaps for Harris, this turn to televising himself can be read as an enactment with his beloved, his watcher, the TV.[54]

But let's return to the 1960s. Surveillance and television became yoked for Harris at some point in childhood, not just creatively but psychically.

Even if you're not a Freudian, it doesn't take a lot of work to see where these wires might have crossed—with his father working for the CIA, as a watcher, and his mother caring for others. His parents were, internally as well as materially, replaced with television. Or parental attention got commingled with mediated attention, parental presence became media presence. Television was the maternal environment or, as I argue in this chapter, conceived of as a maternal proxy not just psychosocially and in panic, but internally. Harris turned watching TV—an act of consumption—into being *watched* by TV. TV performed this basic care function. His life's work culminated in finally getting his TV-mother to actively mind him. Watching was now the basis for a two-way relationship in fantasy (as Bertolt Brecht would have it, writing about the radio, it would be the "finest possible communication apparatus in public life . . . if it knew how to receive as well as transmit, to let the listener seek as well as hear, and bring him into a relationship instead of isolating him").[55]

By the 1960s, on the grounds of television watching in the absence of mother, the media-as-mother panic had begun to take shape. Harris, unwittingly, was its subject and can be seen as anecdotal (if not extreme) evidence of the latchkey child and their television. Television reached children at home, and so recalled the nanny panic, but in the 1960s, mother was increasingly unlikely to be there to co-watch. There would be no mother medium to interfere with media-as-mother. If television was the "boob tube" or an "electronic babysitter" or even a "member of the family," it was clear that in the circuit it created with the child, it was mother-like. But by the 1960s, it was no longer mother-like; it was understood *as* a mother-proxy, and thus threatening to her, to her child, and to society.

Harris may have had an exceptional life, but his relationship with the television conforms to the panics that began to erupt around it. It might be said, too, that Harris, or his relationship with the television, reverses the scenario of "It's 10 pm, do you know where your children are?" to "It's 10 pm, do you know where your mother is?" "Whither the mother?" was exactly the question in this hinge decade of middle-class women's employment and its meanings: with the mother absent, a new focus was put on her adjunctive forms of care—namely, the television.

Technology has a multiplicity of meanings in the home: it is in opposition to a mother's undiluted love, requisite for her labor, a site of panic, and a space of betterment for the child. The opposition between technology and mothering is a screen for the opposition between work and the special case of maternal care. The unwaged labor of white middle- and upper-class mothering was seen as a kind of purity starting in the nineteenth century. The very mothers who *have* to work are considered unfit for having to do so; the fact of work is therefore itself a screen for racialized and classed understandings of fit mothering. It is these same mothers who might serve as proxy mothers, other mothers, or maternals in the role of nurse or nanny. This is because, by the 1960s, screen media, more than any other, were understood both to be proxy mothers and to undermine mothering.

Historically, that work may be waged and elsewhere, or it may be waged piecework taken into the home, or unwaged labor to reproduce the home, just a room (or a few feet) away (see the next chapter). Whereas it is a cultural fantasy that women, and especially white women, did not perform waged work (let alone out of the home) until the 1970s, this is not factually true. In parallel to calling on extended kinship networks for care and the employment of paid caregivers such as nannies (depending on class position), children also took care of themselves—on their blocks, with their friends, by their siblings. They were not alone.

As we have seen across this book, in addition to peer relations and physical architectures such as cribs and playpens, media were and are called in as containers—using attention instead of presence to contain even as they do not confine a child's body. Whether the optical toys of the 1890s or the television screens of the 1960s, media were recruited to direct the body to purpose or passivity, whether via entertainment or the remediation of work, or some admixture of both.

Across the 1950s and 1960s, the psychological literatures on mothering show time and again that when mother is home, she is supposed to be *home*—not just physically but attentively. This is in part because, as mothering intensifies in the postwar era through our present, mother becomes conflated with the home—she becomes the total environment. Indeed, as Spigel writes, she is commanded to presence *and* attention. She is not supposed to substitute herself. From the 1950s onward, when mother fails to appear attentive and present, her very fitness as a mother is questioned: at the doctor's office, by peers, and even perhaps by herself. What one mother is in danger of "passing on" to her child is quite different from another, located in identity. But when mother was replaced with media, suddenly the limits of broadcast and the medium specificities of television made it a proxy-mother where all etiological outcomes were presumed to be the same. Television offers, following communications scholar Paddy Scannell, an "anyone-as-someone" communication structure.[56] Instead of the particularity of a specific mother-child bond or maternal impression, with television presiding as batch-mother, children were supposedly left open to what flitted across the screen.

RERUNS

In the mid- and late 1960s, the panic against television had solidified on the grounds that children were too porous to its content. Even as the comic book panic receded, illustrated children's books, including those by Maurice Sendak, continued to be charged with brainwashing and mind control by psychologists (Bruno Bettelheim, who treated "Joey," was particularly enraged by Sendak's *Where the Wild Things Are*). Yes, television might keep children off the street, but, as with comic books some twenty years previously, many people were now staunchly subscribing to the idea that television corroded youth—from sexual panics to panics about physiology. Again, as Wertham had described sixteen years prior, a major worry was a psychological one, that television taught children mimetically, and specifically about violence. Much like theories of the violent family raising violent children, television was now purportedly the instrument that, regardless of family life, turned every child "uniformly" violent.

Policymaking around television and its child had seen a sea change in less than two decades. In 1950, as part of the White House Conference on Children and Youth, many of the signal, if not unwitting, architects of the mother-as-medium idiom—Erik Erikson, Dr. Spock, James Clark Moloney, and David Levy—were asked to make recommendations for societal-level interventions for the betterment of children. They *recommended* investing in television for children. By 1969, things were altogether different. Senator John Pastore moved to commission the surgeon general to investigate TV violence as a public health problem. Three years, and $1.8 million later, the report debuted to much confusion over its call for "immediate remedial action." What that action might be and who would perform it (families? broadcasters?) was unclear not just to the American public but to the very experts who had testified to the Senate subcommittee.[57] The mimesis hypothesis was understood as an effect of the medium, but the warning was not modeled on earlier media panics alone. The Office of the Surgeon General made, and makes, very few public recommendations. Most recently, it had taken on smoking and its effects on smokers. Here, television, and its effects on watchers, appeared under this same rubric of public health crisis.

It was against this backdrop that children's educational programming was coming into its own after a decade of initial programming. Two suppositions had to be made for its premise—that television could function as education—to work. First: that media could pacify children, keep them still. And second: that media could teach—which had been an early finding of the Payne Fund Studies. If media were absorptive and addictive, then not only would they keep children off the street, but the medium's affordances would allow messages to be imported, broadcast, disseminated. This meant that a set of new questions could be posed: If television could miseducate, could it also be harnessed for good? If informal education was de facto what a film did, could *formal* education exist on television?

As media historian David Buckingham writes of children's television as it emerged in the United States in the late 1950s and 1960s, it was "not produced *by* children but *for* them"; we can see the fantasies inherent in producing media for a group that had—at least initially—no share but a deep stake.[58] Early TV programming for children was quite marginal, produced to be watched by the family together—especially a child with its mother (in Britain, across the 1950s, the BBC's preschool programming was called *Watch with Mother*, teaching viewers how to use the program in its very title; other key shows in this period were narrated by a "mother" for mother and child).[59] In the United States, children-specific programming was largely known as the Saturday morning cartoons, and in the 1960s had mostly become thirty-minute animated syndicated shows—in essence, moving comic books.

Sesame Street's makers—namely, its television producer Joan Ganz Cooney and Lloyd Morrisett, a psychologist and the vice president of the Carnegie Foundation—put plans in motion to create a preschool-aged television show not in spite of the perceived problems of television, but through them; as Michael Davis writes, their earliest aim was to "master the addictive qualities of television and do something good with them."[60] The show was to draw doubly on the logics of public health and education, with an eye to redressing not only the paucity of children's broadcasting (reaching the little children watching test patterns, snow, and cartoons), but the wider inequality in education in the wider social world. The show received funding from the federal Head Start program, which, beginning in 1965, attempted to close the achievement gap of children without preschool by offering not just education but a whole host of social services.

The show was astoundingly well-researched—child psychologists held research seminars to take on questions of pacing and preschool attention spans, resulting in the show's signature segmentation (which also mimicked feminized TV of the 1960s, which had lots of repetition around advertisements, designed for a woman doing work while listening, walking from room to room). Psychologists informed what would appear on *Sesame Street*'s screen, not just its rhythms. In the seminars that moved from cognitive processes to emotional effects, many gathered to debate the leading research on television (Maurice Sendak was even sent for and doodled throughout the meetings).[61]

Whereas the telephone in this moment held similarly contradictory meanings—some thought it was responsible for suburban alienation and obscenity, others for the mixing of classes and races and broader experimental modes of coming together—the television was *more* ingrained in daily rituals at this point and, stunningly, more available within the home. Television, with its from-someone-to-everyone structure, could be read, erroneously, simplistically. It was truly a mass domestic medium. Suddenly, television was posited as a great instrument of effecting equality (the other side of producing mass delinquency). Yes, living rooms and families varied; the television did not. The medium could be made good by a good message.

Television was a tightly regulated medium, and unlike other media, say, the radio, audiences could not directly intervene on what they heard (or saw). Home reactions of course varied, and users did work to interact with their televisions and the space around them,[62] but they could not change what was on the screen. There was no early amateur television. Given that television was understood to be so powerful as to implant its message directly in the mind of its viewer—for *Sesame Street*, one understood to be a preschooler—it could obviate difference, so long as it was present. With the increase in attention to early childhood education—as well as childhood development (see the preceding chapters), the standardization of the child rhymed with television's standardized form of communication and a uniform vulnerability to its messages.

One participant asked to advise *Sesame Street* in pre-production, Dr. Chester Pierce, was a founding member of Black Psychiatrists of America, and became a key consultant on the formulation of the Children's Workshop and *Sesame Street*'s resulting programs (which had Black authority figures, including a teacher, and would soon increasingly diversify). In the second seminar held by *Sesame Street*'s founders, Pierce spoke directly to the largely white group, saying in essence that they were out of touch with the reality of the very children they were trying to reach: "I listened to this all morning and the image of the show that was building in my mind was absolutely horrible . . . [how could it do] anything other than destroy the viewer?"[63] Pierce, who is now most famous for coining the term "microaggression," heartily subscribed to the idea that media and racial violence had something to do with each other and that representations could be, in essence, annihilative.[64] He put the goal of *Sesame Street*, in the second seminar, bluntly: he wanted the show to prepare his daughter for racial violence and derogatory attitudes—in short, the violence of growing up Black in America.

Pierce attributed child development to a much more complex range of factors, those that extended beyond the family dynamic (family therapy, as historian Deborah Weinstein has shown, had come to prominence by the 1960s) to include the social in the psychosocial. As we saw with Wertham, this was part of a minor tendency in the psy-fields. Racial violence and anti-Blackness were understood by some—but not those in the halls of mainstream psychology—as key concerns for child development. Where this was understood, it was often to disastrous effect. The Moynihan Report (see chapter 3) had just condemned the Black family for, in essence, not being one—for being dyadic, where Black children had no one else to *mediate* their mothers. Simply put, the Moynihan Report described the absence of the Black father in the United States as a crisis of under-mediation of the mother, with no one to de-Oedipalize the Black child. Whereas many media panics that centered on children assumed the reader or watcher to be white, recalling the racialization of the innocent, priceless child, Pierce knew—as all astute psychologists interested in media of this moment did—that television was a *near* universal, as 95 percent of homes had it by the time *Sesame Street* debuted. And he knew it intimately: he had a three-year-old daughter.

What happened to any given child in front of the television was therefore, similarly, a universal experience on this material understanding: almost all kids had TV, almost all kids would experience it the same way. But Pierce was particularly concerned about what happened when Black children were exposed to television programming made for the universalized white child. TV programming was made for "everyone" and all children were television's children, made porous by television's hot someone-as-anyone structure. But whiteness is not universal, even as psychologists in this period took the white psyche to be so. While televisions were universal, how children interacted with them would have mixed outcomes based on their own lived experience (let alone their fantasy lives). Pierce wanted the producers of *Sesame Street* to understand that, even if their medium could reach *everyone*, television's child was not universal; nonetheless, TV was covertly offering a white for-someone-as-anyone structure of communication. Pierce's interest in what *Sesame Street* proposed was that it could go beyond harnessing the addictive qualities of television to use diversity of representation to speak to more someones—including his daughter.

By the time of *Sesame Street*'s debut, about two thirds of US televisions were able to pick up the show. Many readers might assume that *Sesame Street* was deeply well received given its longevity. By one token, this is true: even those suspicious of television as a medium of care or education for children wrote highly of it. This included Dr. Spock, who thought that a crop of preschoolers raised on the show would give the nation "better-trained citizens, fewer unemployables in the next generation, fewer people on welfare, and smaller jail populations."[65] Les Brown, writing in *Variety* after the show's debut in 1969, wrote, "Not until the closing weeks of 1969 did television offer a program series that really answered the long-standing criticism of the medium—namely that it takes a viewer's time without giving anything in return—and held out hope for a more substantive future."[66] The television was now seen as a good mother, responsive to her child. It was also seen as economically effective, and drew comparisons to Head Start, now four years old, especially in the first year of the Nixon administration, which had inherited the program. President Nixon's early education officials purportedly said, "We can get *Sesame Street* to reach poor kids by spending sixty-five cents per child . . . why should we spend over a thousand dollars per child on Head Start?"[67]

Whereas Spock thought of the power of educational TV in flat, if not racialized, terms—that good shows would change the citizenry—more nuanced criticism of the program began immediately, especially from leading childhood education and child development psychologists. (There was also racist repudiation of the show, which integrated television preschool; in 1970 Mississippi banned the program, only to reverse course after much outcry.) This included those intimately involved with Head Start. Founder of Head Start Urie Bronfenbrenner and Frank Garfunkel, who directed the Head Start Evaluation and Research Center, each were *Sesame Street* detractors—and for a range of reasons. Bronfenbrenner found, despite Pierce's interventions in the planning and research seminars, that the show "disappeared through a manhole" all the actual pain of childhood.[68] Garfunkel called it a "mirage" and intimated, via the imagery of lynching, that it strangled children.[69] Psychologist Leon Eisenberg, who was responsible for the longitudinal follow-up studies on Kanner's original group of autistic children, similarly concurred on the grounds that the show did not reflect the difficulties of childhood and, for all its interest in social justice, took a "tokenistic" approach.[70] They joined a chorus of Head Start supporters who might have been intimately aware that *Sesame Street* would be seen as an easier, cheaper intervention for children. Head Start would now be under threat in the Nixon administration. Head Start indeed pulled its funding of the show, deeply aware that the tele-education provided by *Sesame Street* might be more appealing than the nation-wide program and its costs.[71]

Mainstream commentators followed the publicity claims that the show was good for all kids, while redressing economic disparity in education. Yet according to critic Michael Davis, there was already dissent in newspapers, if minimal. Arnold Arnold, writing in the small *Hackensack Record*, responded to the very first season of *Sesame Street*, first concretizing what would become a major charge against all educative media in the decades to follow: "It has created unfounded hopes for improving the education of poverty children and may be harmful to them."[72]

Two years later, in an exposé for *New York Magazine*, after multiple seasons had aired, Linda Francke wrote that the show was predicated "more on faith than on evidence" and that, anyway, the same Black viewers the show hoped to reach found it "phony" or outright offensive.[73] Francke quoted the staff of West 80th Street Day Care Center as taking issue with the racialization of the Muppets—even if the humans were diverse and integrated—especially Oscar the Grouch. Director Grace Richard was quoted as saying, "I react to the garbage-can character because that to me is the inner-city character. He's the one who's bottled up, and who compensates for it by saying he likes to live in a garbage can. That's really like saying it's all right to live in a dump." This sentiment was echoed by Dorothy Pitman Hughes: "That cat who lives in the garbage can . . . should be out demonstrating and turning over every institution, even *Sesame Street*, to get out of it."[74] Many who were interviewed for the story reported that *Sesame Street*'s representations meant to address the inner-city child instructed them, perhaps unwittingly, to consign themselves to a supposedly predestined fate.

While *Sesame Street* was litigated in the press (the show won), psychologists continued their work on the show, studying its first two seasons. Psychologist Herbert A. Springle offered a two-pronged approach to their critical studies, as Robert W. Morrow writes, "pedagogical and political."[75] They were particularly keen to understand if the show met its own aim, that is, to close the achievement gap. Some startling clinical findings began to appear. Yes, the show closed the achievement gap in the markets where it was watched, but primarily when children *watched with their mothers*. Recalling the notion that the addictive, pacifying elements of the television medium could be used for good, *Sesame Street*'s producers had long rejected the claim that children learn only dyadically—that is, they cannot acquire preschool-level concepts passively. Springle argued that they were, in essence, deluded. While children of color and those living in poverty who saw the show had their IQ scores raised and their emotional/social learning increased, the white, middle-class children's IQs went up much further. Moreover, children who already had access to preschool fared much better than those who watched *Sesame Street* without a school component. In essence, the gap stayed the same or widened.

In 1971, Thomas D. Cook of Northwestern also began his investigations into the show. Cook sought to isolate the impact of *Sesame Street* on learning. What he found was that those in "encouraged families"—that is, families that had social services come to teach them about how to use the show as part of the study—and who then watched the show together saw achievement accelerate in their children. Children who watched alone did not see any gain. Cook argued that *Sesame Street* and educators in the home who taught parents how to use it provided material for parents to teach children—but the show alone, isolated on the television screen with an unaccompanied child, did not. While educational television was born from the "wasteland" of programming, it was also born of the understanding that children who watch television alone do so because mother is elsewhere: with other children, out of the house at work, doing other reproductive labor. On Springle's and Cook's understandings, privilege was thus defined by having both mother and media at once. Mother-as-medium and medium-as-mother needed to work together, but when mother was absent and media were truly doing the work of standing in for her, children would not flourish *as much* or as quickly. The many children *Sesame Street* hoped to reach sometimes did not have televisions or (televisions receiving *Sesame Street*) to begin with. If they did, it was also assumed they would have media at home during viewing hours, but not mother. If the assumption that television was a medium beyond reproach lay in its social context, namely, that it brought the family together, and more specifically that mother would be on hand—her presence if not her attention—while child consumed media, then that assumption was upended doubly by *Sesame Street*.

MEDIA, METAPHORS, MOTHER

If *Sesame Street* was largely preserved in its public reception across the next fifteen years, television as a medium—an intruder in the domestic attention economy—became a more intense site of scrutiny. As with earlier studies, there was still a fixation on *the media*—that is, the content on TV, as in the classic example of George Gerbner's 1968 "mean world hypothesis," which argued that violent media warped our perceptions of how violent the world was. But *the medium*, or just the fact of TV, irrespective of content, was increasingly the site of panic. Immediately in the aftermath of *Sesame Street*'s debut, new popular books worked to explain the impacts of children's television: in 1971, Norman Morris's *Television's Child* appeared, solidifying the metaphor of a child being mothered by an appliance, while William Melody's 1973 *Children's TV* tried to explain the relation between the hostage child consumer and the young zombie before the flickering light that appeared in the living room. Soon, shelves of economic, psychological, and political anti-television treatises appeared; many were superficially disconnected from the 1950s panics about television. The language and methods were no longer focused on "bug-eyed" children but on attention spans, less concerned about mimetic violence than deeply invested in "behavioral issues." Despite an entire generation having now absorbed the medium, the critiques remained, in a way, largely static, as had viewing numbers.

Television criticism reached saturation point in the mid- to late 1970s. Upon reviewing Marie Winn's 1977 *The Plug-in Drug: Television, Children, and the Family*, the *New York Times* critic made the following wry comment: "having someone tell us one more time that television is the opiate of the people is about as startling as confirmation from the White House

Press Office that Jimmy Carter brushes three times a day."[76] Winn's book, which was published concurrently with Jerry Mander's manifesto in favor of the abolition of television, *Four Arguments for the Elimination of Television*, regressed the television critique from content-specific debate to medium-specific debate. In the aftermath of *Sesame Street*, the new television critics of the 1970s felt they could confidently mount a more total indictment of television: it did not matter how good the show was; the medium was, in McLuhan's terms, the message—and the medium was bad to its core.

McLuhan's moral and ethical media theory was taken up further across this period by the second generation he had trained, who now called themselves media ecologists, following Neil Postman. In 1971, Postman founded the Department of Media Ecology at New York University (the department where, some forty-seven years later, I would earn my PhD, although the name and commitments had been rather reshaped in the intervening forty years). Postman developed McLuhan's "medium is the message," rerendering it as "the medium is the metaphor" and, across the 1970s and 1980s, took aim at the metaphors of television and its impacts in two books, *The Disappearance of Childhood* (1982) and the equally sanguinely titled *Amusing Ourselves to Death* (1985). Postman here extended theorization of media effects—arguing that media, like a mother, are responsible for not only the creation of childhood (where print cultures split children from adults) but its end—when adult and child content merged on the screen. Television's child, it turned out, in addition to no longer having a human mother, was no child at all. There were no children, not really. Even as the dominant metaphor of the television-child relation continued to concretize a child's dependency, passivity, and capacity for absorption, media theorists were decrying that video killed childhood.

Unlike the panics surrounding mother-as-medium, in which the particularities of each mother-child dyad were classified, at scale, the scale of the television now classified children. Or as the media critic George W. S. Trow wrote in 1980, appraising the cultural shift wrought by television in the preceding two decades:

> The middle distance fell away, so the grids (from small to large) that had supported the middle distance fell into disuse and ceased to be understand-

able. Two grids remained. The grid of two hundred million and the grid of intimacy. Everything else fell into disuse. There was a national life—a *shimmer* of national life—and intimate life. The distance between these two grids was very great. The distance was very frightening. People did not want to measure it. People began to lose a sense of what distance was and of what the usefulness of distance might be.

This contention—that daily life was now split in twain across two sectors, the grid of the national and the grid of the familial-sentimental—was consecrated and literalized in the space in front of the television. The only remaining distance was that between viewer's nose and the screen. When the panic of television is told, it is often told as an addition—and of course it is that too. But Trow marks also a deletion—of the middle distance or of society itself. The two remaining grids—so to speak—were that of the family as media, and media as family. Television becomes society—becomes the place in which to intervene. And it enters—as intervention—the family. The metaphor of mother-as-medium had been fully recapitulated as media-as-mother. Decades before the iPad appeared and was converted to soothing device, media-as-proxy mother was found not to supplant—media could not mind, could not pay attention. But children could pay attention to media and treat media—in what would become known in 1996 as the "media equation"—as human. Meanwhile, media theorists and psychologists alike maintained that media and mother have the same ferocious effects on the child.[77] The maternalization of media was complete. As we will see in the next chapter, the deletion of middle distance, of society, was only further reified when mother and media merged into an all-encompassing environment: the home.

6 FUTURE, TOMORROW, DREAM, SMART

Asked for a comment on the impeachment of the president of South Korea, Professor Robert Kelly appeared on BBC Live in 2017. As he began to speak, first one young child, then a second, burst onto the screen in his home office. They were quickly followed by their mother, Jung-a Kim, who frantically pulled the children out of their father's remote television studio. The video quickly went viral and was viewed hundreds of millions of times. It was cherished for its inadvertently comic timing but was subject to (mis)interpretations that leapt from the Internet into the family's real life: commenters sympathized with Kelly, shared how often a similar situation had happened to them, chided his wife for her lapse, and in a racist manner miscast Jung-a Kim as a childcare provider (or "the nanny"). The viral reception turned so heated that the family had to use security guards at their home and children's schools for a year. Beyond the vagaries of virality, the video encapsulates both the real labor and fantasies that subtend telework. Kelly's strategies for separating work from home in telework failed: his children did not stay put in their planned activities, and his wife was one step behind them.

The fantasy of remote work contains an inherent contradiction, in which the flexibility of home work allows for the care of family and domestic space and, simultaneously, presumes the home can be bracketed so that productive labor—always via the notion of suspending reproductive labor—can occur. It also occludes the multiplicity of workers and kinds of labor that can only be performed in person (including childcare itself). The idealized notion of remote work as solution to the double bind of caregiving workers is breached when these two spaces overtly collide; to keep the fantasy going, children must be kept out of the frame (although the pandemic has relaxed the norms and feeling rules around this question somewhat, for some workers). Once children are in the frame, the domestic and the office hopelessly clash in remote work (and can damage careers and become sites of quotidian shame, let alone internet stardom). Kelly was able to rely on a home office, and on a partner to perform childcare, even if these strategies momentarily failed to keep his kids off-screen. What of those workers without childcare? How do they manage to preserve an extra-domestic work environment within the home? In this chapter, I will look the history of "smart homes" and futural architectures that have attempted to allow mother-as-worker and mother-as-mother to meld with house-as-mother. The home as a media-mother is the final extension of mother-as-environment, becoming environment as mother and returns us to the very scene where we began.

The history and conceptualization of remote work—itself usually feminized—implies an impossible disarticulation of space in which one is neither at work nor at home. Much has been made about the loss of the "third space" (the bowling alley, the bar, the library) in COVID-19 life, but many have also lost a distinction—insofar as it existed—between work and home, two spaces becoming one. In parallel, in the decades-long history of telework, dislocating labor from an office has been forecast as a necessary future and a site of liberation. Working from home and eschewing the office has been heralded as an ecological necessity, good for individual wellness and for corporate productivity. Telework has additionally particularly addressed itself to women via—and despite—this contradiction of being nowhere and somewhere at once and has relied on the notion of a feminized "having it all" and the "flexibility" necessary to accomplish it, to collapse waged labor and domestic work, where having it all implies some form of difficult or impossible simultaneity, being everything at once and nowhere at once in order to have everything.

To support this feminized fantasy of a freer future, with neither the full drudgery of office nor the domination of housework, domestic architectures and infrastructures offline have supported the possibilities of teleworking: rooms are configured to dampen sonic leak, the increase in children's homework and after-schooling allows some separation from domestic labor and child, and the home office evolves as a distinct, technologized space, to which domestic computing is introduced.[1]

Telework has been coded unevenly into the very artifice of middle-class homelife since the 1960s. These traditional physical architectures have addressed themselves to the challenge of containing the chaos of the domestic and readying individuals for productivity and work.[2] Women faced with the realities of working from home while raising children have seen, across this history, that these physical separations are never enough. Like Professor Kelly, these workers have seen that, *pace* Virginia Woolf, a room of one's own is not enough—if one even has that. Professor Kelly's viral video lets us see the a priori conditions of remote work: it renders home and office indistinct without collapsing the separate and antagonistic roles of the person involved in those spaces and activities. In order to work from home, one must first disappear home from work. To make use of the supposed flexibility in caring for children that telework promises, the children must not fully appear or exist on the clock.

If, as the feminist critic Sophie Lewis writes, the "private household is the dominant psychological architecture of modern living," this specific exclusion of remediated childcare in smart homes reifies and contributes to the production of an ideally unmediated (and biological, and white) motherhood.[3] It is architectural evidence of the mystification that, Silvia Federici (among other socialist feminists) argues, has allowed the family to function as "social factory."[4] Theorists of feminist technoscience have moved beyond describing this mystification to look instead at how technology, gender, race, and class produce these partial realities of the social factory. Parenting technologies and their ancestors have long been part of the assemblage of that factory, and across the twentieth century they take on the status of surrogate humans.[5]

In this book's first two chapters, on remediations of two of the dominating gestures of childcare—watching babies and getting them to sleep—we saw the use of cribs, monitors, and screens to extend the reach of care, or the media of mothering. In the book's middle two chapters, we saw mothers turned into media, into inputs and outputs. Now, in the last third of the book, we are looking at how media were understood to pick up the function *of* the mother. If, in the preceding chapter, we saw how media were conceived as mother in her *absence*—when work took her away—we will here look at the collapse (and fantastic replacement) of mother and media in the name of waged labor.

An adage of histories of technology is that a feminized technology is one that can be picked up and put down in exchange for a child. Telework sets a condition in which all home work becomes feminized work. Time-saving technologies—since the turn of the last century—have been escorted by the claim of easement: that we might lessen domestic burdens and increase environmental control. As Cowan has classically shown, this sleight of hand merely increases unwaged work. This chapter investigates the collapse of that unwaged work with the increasing demand to be always on—both as mother and as waged worker. Yet the very first forms to make this gesture, now so commonplace, were minor or, in their inventor's own words "less than epic."

HOME, ELECTRIC

The modern features of the automated-home-as-mother were compressed and forecast in its earliest example. The development occurred in tandem with the electrification of urban households and the rapid onset of the decline of domestic service. The first proto-smart home—a total system for mitigating certain types of housework—appeared in earnest with General Electric's Home Electrical, which was mounted at GE's exhibit at the Manufacturer's Palace for the Panama-Pacific International Exposition in 1915 in San Francisco. While attempts to partially automate women's social reproduction, their home work, began in the mid-nineteenth century,[6] GE was the first company to sponsor the making of a model home featuring innovation in *almost* every conceivable space (mostly focused on heating, cooling, and lighting, but also stand-alone appliances such as a vacuum cleaner, sewing machine, and cigar lighter—all refashioned by GE).

In doing so, General Electric translated the feminist critique of its moment—that hyper-focus on the wrong domestic tasks ruins a family—into an aspirational consumerism that maintained mothers in the home, with a shifted emphasis on their role. This argument is legible everywhere in the design and advertising for smart homes from the turn of the twentieth century to the turn of twenty-first. In 1915, less than a third of American households had electricity, and the Home Electrical was not only a model home as display case for GE devices, but a showcase for domestic productivity and a house-sized advertisement for electricity itself.[7] To advertise the advertisement, GE made a short promotional film, "The Home Electrical" (1915), which won two prizes at the San Francisco World Fair that same year.[8] The film was continuous with GE's earlier attempts to advertise domestic appliances to audiences via short one-reel films coupled with giveaway prizes for filmgoers as early as 1911 with *Every Husband's Opportunity*, commissioned by GE and produced by the Essanay Motion Picture Company.[9] As John W. Schwem of GE's Apparatus Sales Division recollected some forty years later, this promotion may have been the first industry-sponsored film.

The film was "something less than epic": it opened with a husband and wife distressed about their domestic situation and, importantly, frustrated with their human help: their servant. The couple is shown to be much more harmonious once receiving electric aid in the form of GE appliances.[10] GE's Schwem noted that reception was mixed: theater owners found *Every Husband's Opportunity* too on the nose—more advertising than story—and projecting 35 mm was difficult if not outright prohibitive.[11] But where the film did screen and prizes were given, house-wiring contracts followed.[12]

By 1915, GE had its own in-house film production unit, headed by Schwem. In *The Home Electrical*, viewers follow along as a visitor goes through the home with his wife and friend, with intertitles that pose rhetorical questions about the feasibility and function of all the new splendors of an electric house. The kitchen was "the pride of the house," touted in the promotional film as envy-inducing, time-saving, and astonishing in its advancements (kitchens have largely been the easiest site of innovation in technologized homes). It is there to "assist" the housewife in her "toil."[13] After dinner is served, the wife excuses herself to check on her baby. It turns out that "Billy" is no baby at all and does not really need to be checked on: he is both school-aged and tended by a nanny (an older white woman) in uniform. The inclusion of the nanny allows the viewer to ascertain that mother is not left to mother alone in this house filled with technological aids; the assistance in mothering is still human, not electric (see chapter 1). Despite the inclusion of the human nanny, the mother writes a note—addressed to General Electric itself—thanking the Home Electrical for bettering Billy, who is much improved (what was wrong is never stated)—thanks to his electrical devices.[14] In this, the advertisement asserts that the supposedly neutral care via technological aids helps the child, occluding both human mother and nursemaid as nurturing inputs.

In these consumer logics, time-saving devices in the home have ostensibly carved out a space for mother. Following Cowan's famous axiom, time-saving in the case of domestic drudgery in turn produces *more work for mother*. Where Cowan focuses on the extension of labor under the guise of reducing it, we know from the history of the nanny that when childcare is directly addressed by time-saving it is most frequently to remediate a *human helper*. Additionally, these devices often create more work for mother when she is called to manage the surrogate humans—whether via fear, hours logged reviewing video, or charging and maintaining devices. As the Home Electrical continues to show, the labor remediated by "time-saving" is often a fallacy, but nonetheless a persistent one. After all, as Sianne Ngai shows, time-saving is nothing if not a gimmick.[15]

THE HOME OF TOMORROW

Across the twentieth century, something else happened in the electrified homes of the future. As the status of work shifted in the middle-class home, a strange *return* took place in the case of married women taking in work, what had formerly been known as piecework (and what I will call home work). As historian Claudia Goldin notes, at the turn of the last century, fewer than 10 percent of married women worked. By 1920, that had changed: one in four married women now were employed.[16] Repressing the history of women's work, the push for telework appears as if new, as an innovative way to manage the burdens of being both productive and reproductive. As we will see, as the home becomes networked, the ability for mother to be a housewife, to mother, and to productively labor all at once becomes the goal.

After GE's enterprising Home Electrical, staged model homes continued to introduce the public to new forms of technologized and automated living that frequently, though not exclusively, ameliorated the toil and duty of womanhood—but which toil could or should be alleviated continued to be a site of bias. The genre of the show home, or model home, is a form of middle-class entertainment[17] that works two ways: first, it is a completed object that tries to sell a particular form of house, staged to give a thick sense of the life one could lead there. "Model" means something else too—the model home elaborates an archetypal life, one that can be purchased and imitated via interaction with architectural features. A model home is, after all, only a model, a prescriptive space, a predictive space, a nostalgic space, and thus an ideological architecture.[18] In this category of architectural project, the future was frequently a techno-solutionist and optimistic version of a past that was, in Stephanie Coontz' apt formulation of the history of the family, "the way we never were."[19]

In 1933, in the middle of the Great Depression, eleven such model homes were installed at the Chicago World's Fair. The homes pointed to a future filled with plenty. Each home was sponsored by architectural firms, corporations, and magazines (*Good Housekeeping* had its own home in the pavilion).[20] One designed by George Fred Keck explicitly took up the title of "The Home of Tomorrow."[21] The upper two floors were all glass, the kitchen featured GE's first dishwasher and Goodyear rubber floors, and the structure introduced consumers to amenities that would shortly become standard in many homes: the attached garage, air conditioning, forced-air heating, and automated doors with sensors.[22] While other exhibits at the fair featured celebrations for groundbreaking technologies in pediatrics, these homes did not communicate an overt interest in creatively redesigning the living quarters of young children or innovating how parenting would be done in the home.[23] From its location in the 1930s, this house occupies an interstice between live-in care and a dedicated playroom for children as the solution to child containment.

Just six years later, at yet another "Town of Tomorrow" showcase for the 1939 World's Fair in Flushing, Queens, architectural innovation contained children via play directly: Landefeld & Hatch included a playroom in their model house that would, within the decade, become an ordinary architectural inclusion for middle-class homes, supplanting the notion of a nursery (in which children most frequently did not play).[24] As historian David Snyder shows, this room for children's particular use facilitated children's self-amusement and kept them from interfering with their parent's leisure needs across the 1940s and into the postwar era.[25] Media historian Elizabeth Patton argues that, at mid-century, a space dedicated particularly to children and their use "was framed as necessary for developing a sense of responsibility, security, and 'pride of ownership.'"[26] Separate spaces for children were not altogether new, even if offering specific rooms dedicated to their leisure might be.[27] While this might have the intended benefit of preserving space and time in the home for parents while children were near, the distance between rooms also introduces a fraught anxiety: if children are out of sight or earshot, they are not being minded and are being denied maternal or parental presence and attention.

AT HOME WITHOUT CHILDCARE IN TOMORROWLAND

Time-saving devices were not all treated equally where they addressed childcare rather than chores. The crib, by one conception, is a time-saving device, as is an infant's schedule, but it was not named as such. Instead the "very fine" labor-saving devices of the postwar period were put to work in the name of ending domestic drudgery but not childcare, and were increasingly in reach of the middle class, as were new forms of housing to hold them.[28] As part of the GI Bill, white veterans could make use of a whole host of benefits upon their return from service, including help with securing mortgages.[29] This wave of government-sponsored home ownership intersected with the legacy of the Great Depression and World War II construction restrictions to produce a massive housing shortage.[30] The lack of homes for sale was an architectural and engineering problem that produced an industry of home fabrication at a speed that could match demand. By 1958, 10 percent of Americans lived in prefabricated buildings, predominantly in suburbia.[31] At the same moment, there was already a burgeoning electronics and appliance industry concerned to automate the home and especially the domestic labor that occurred within it. The rise of the prefabricated postwar suburban home and its new appliances was meant to lessen the load on mid-century housewives at an attainable price, especially as the young women of this same group continued to conceive the many children of the "baby boom"—the same women Skinner and his sterile crib had hoped to help with child-rearing, just as they'd been "helped" with other forms of domestic toil (see chapter two).[32]

The 1957 Monsanto House of the Future at Disneyland's Tomorrowland offered a different form of speculative architecture, showcasing the future of domestic work via its interior and its sophisticated use of Monsanto plastics.[33] The House of the Future took increasingly commonplace features of domestic life and put them through a Cold War time warp.[34] The house was funded by and fabricated by Monsanto, supported by Walt Disney Imagineering, and designed by MIT architects Marvin Goody and Richard Hamilton, as well as other MIT faculty. The house was "set in the future" (the year 1986, like the rest of Tomorrowland) as a site for women to enjoy; the rest of Tomorrowland was filled with masculine attractions.[35] The home was staged like a contemporary purchasable prefab model that invited "prospective buyers" to view a product that was not actually for sale, in that it was demonstrably unattainable and not meant to be produced. Between 5,000 and 10,000 visitors poured through the house on any given day to ogle the "dream of the future, brought alive by Monsanto," a futuristic, sci-fi, quasi-Googie home filled with "wonders": lights that were on a dimmer, intercoms, ultrasonic dishwashers, "cold zones" instead of refrigeration and freezers, climate control (not just heating and cooling, but atmospheric salty sea or woody pine air piped into any room in the house), plastic floors that were easily cleaned.[36] (Two of these visitors include my grandmother and mother). Dad is shown using what may be the first instance of a closed-circuit television system in a private residence; he checks the front door while using his electric razor in the bathroom, keeping the Home of the Future and, by proxy, his children's tomorrow, safe.

Vigilance and security are displaced as driving concerns by the notion that the home was enjoyable precisely because it offers up so many forms of automated control. The advertisement for the home claims simply that

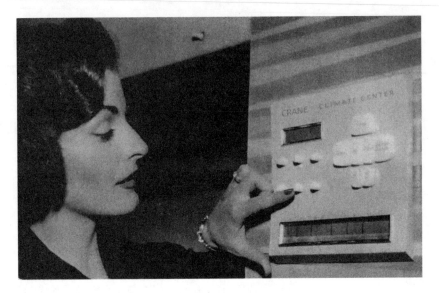

Figure 6.1
Climate control center in the Monsanto House of the Future. Cartwright Collection.

domestic labor is "fun—even setting the table!"[37] The language of joy and pleasure is used to re-describe domestic labor as well as to obscure the house's technological limits in abolishing that labor: "This kitchen almost gets dinner itself—but that wouldn't *really* be fun." The television spot aims contradictorily to make the home seem available while repeatedly delighting in this "pure fantasy" rendered in plastic;[38] Sleeping Beauty Castle is shown in the distance, another fictive home out of reach of visitors.[39]

Much like in the Home Electrical and the Home of Tomorrow, in the House of the Future (the shift from home to house, or a dwelling without present dwellers, is an implicit marker of the model home as aspirational and ideological entertainment, never to be lived in even as it influences desire), the children's rooms are surprisingly underfeatured given that, as David Snyder suggests, "By 1950, child-centric suburbia had produced explicitly child-centric architectural design."[40] But in actual homes of its era the children's spaces were small and central; Amy Ogata argues that quotidian domestic architecture ensured more overlapping domestic space for children and parents.[41] Children's areas frequently abutted the kitchen so that the mother-housewife could perform all her duties at once.[42] In the House of the Future, there are, of course, two children's rooms, one for a boy and one for a girl. The selling point of the son's room, which is rather blank even though staged to appeared inhabited for the promotional materials, is that everything is easily washable, again lessening the toil of domestic labor associated with child-rearing (laundry) without addressing childcare itself. The daughter's room is the mother's segment of the adult bedroom miniaturized. The television spot advertising the attraction deletes all evidence of the son except for his room, while showing the daughter both as a child and a "little mother," helping with all the feminized labor. Even though these children were already on their way to adulthood, marked heavily by gender norms (the daughter tethered to the home, the son ostensibly out in the world), the very architecture of the home never lets its viewer forget that there *are* two children around here somewhere—but we never see the mother *doing* mothering.

There is a technological misdirection here: the architectural and advertising focus on school-aged children continues the smart home tradition of preserving white motherhood as separate from other forms of domestic labor, retaining its human-to-human purity. It has disappeared the school, the bus driver taking children to and fro, the housekeeper. The advertisement conveniently sends the children out of the house or, in the case of the daughter, turns her into an innocent mother-in-training in keeping with understandings of latency and adolescence at mid-century. If child-minding is naturalized as a white woman's domain, despite the long history of racialized childcare, it is a site hypercontracted. It is resistant to remediation precisely because mother has become a medium. We can see this when we compare the advent of child-minding technologies and their reception with mid-century labor automation (whether in the factory or in the home).[43] The latter is triumphantly innovated. The former is considered pathological.

The House of the Future is almost as much fantasy as the scenes of Bradbury's 1950 Happylife Home with which this book opened.[44] Where Bradbury imagined the end of all reliance on a mother of flesh and blood, the House of the Future steers clear of mothering among its assistive devices, and deletes domestic workers from the scene. Whereas intensive mothering is, in our contemporary moment, figured as a moral good, perhaps the avoidance was from uncertainty and warning: at mid-century, *over*-mothering (or momism) was just as much a problem as *under*-mothering (refrigerator moms, supplementing care with television, see chapters three and five). As we have seen, both these states were understood as produced, at least partially, by class and race, technology, presence, and attention. Nonetheless, the fantasy of unaugmented and unmediated motherhood is in keeping with the material, lived reality in domestic architectures and design of the 1950s. As we have encountered previously, the interior spaces and standard equipment of the American working-class and middle-class home went through a sudden, breakneck renovation that rendered the industrialized home commonplace in this period and, as Judy Wajcman shows, these automated devices further reified the long-standing gendered divide around domestic work.[45]

With the increased presence of technology in the home came the trope of the bored housewife, so well "assisted" by appliances and their processes in her daily work that she has nothing left to do but become cold and discontented (not unlike Bradbury's Lydia Hadley). This figure appears everywhere in postwar discourses, in punishing psychological studies against mothers who, despite having nothing to do, do not love or tend to their children "properly." This trope is, of course, at least partially a myth: the housewife may have been bored *with* her work but she was not bored for lack of it.[46] Mothers indeed worked more than ever before, in part due to shifting social structures and in part due to the new, myriad tasks associated with managing their technologized homes. The feminist description of tedium and its resultant affects proved not synonymous with having nothing to do. But the *appearance* of having nothing to do can be remedied: the stove can *almost* get dinner ready on its own (someone still has to shop and pop food in to heat—even if just a TV dinner), while children and child-rearing are left unmediated (even as that dinner *will* be eaten in front of a TV, at least after 1954).

COMPUTING THE HOME

By the time the House of the Future closed in 1967, many suburban homes boasted not just major appliances, but ones with some element of automation such as coffee makers on timers and self-cleaning ovens. Frigidaire sponsored a television short, "A Day in the Life of a Kitchen" (Film Associates, 1965), similar to the advertisement tie-ins for the House of the Future and the Home Electrical (perhaps rhyming with earlier promotion documentaries from the 1940s, *It Happened in the Kitchen* [1941] and *A Step-Saving Kitchen* [1949]).[47] The appliances on display were frequently marketed as leaving more energy for mothering; Westinghouse "Leisure Line" appliances choose to have the corporate line spoken by a child himself, collapsing the aim of the device and child as beneficiary: the tag line "Leave it to me and there will be more time for mothers!" keeps it tactically ambiguous whether that time is for mother or for *mothering*.[48] In "A Day in the Life," we encounter a "typical" kitchen scene, replete with a thriving middle-class couple. Though we watch the housewife go through her actions, it is decidedly a day in the kitchen's life we are supposed to be shadowing. The housewife is reduced to the "operations chief" as well as floor worker, and the proto-smart kitchen has Tayloristic touches throughout to optimize her every move.[49] And yet, managing itself is a new form of labor, as Cowan has shown; no freedom of movement has been granted via automation: the floor manager still cannot depart the kitchen. Her daughter, once more the only child extensively featured, although there is also a son, is old enough to help, be sent out to play, and have a party her mother has curated for her—appropriately outer space–themed, complete with spaceship popsicle molds.

In the same moment, new domestic uses of computing were being imagined and marketed in similar modes as these smart kitchens and the House of the Future. By the mid-1960s, computing evangelists and visionaries were already pushing for home computing—about a decade in advance of the Apple I, with devices such as the 1969 Honeywell Home Kitchen Computer. Additional tactics were innovated to protect the fantasy of being able to do it all, let alone have it all, in the same room. The realities of work from home were managed by women simultaneously performing domestic and paid labor. Beginning in the 1960s, women who worked together at home or were self-employed at their own kitchen table businesses deployed, in one example, tape loops of office sounds and typewriters to transport their tele-clients to the office and cover over the sound of children who were sitting and playing nearby.[50]

Unsurprisingly, these home computing applications—and the sounds set to accompany them—were deeply gendered and were first meant to automate and program the domestic, rather than extend to the productivity of paid work. One representative article trumpeted: "housewives would use the machine to program all their chores most efficiently, prepare dietetically balanced menus, check prices for a particular item at all stores in their neighborhoods, place orders, do their banking, order specific home television shows, and for diversion, attain an advanced degree at a university, all via computer and without leaving the home."[51] The housewife is afforded a telecommuting relationship to her duties outside the home, where she is maintained in its center: the kitchen or family room, or sometimes at a terminal. This version of a digital public sphere rescinded a woman's invitation into public life even while positing her participation in education and the professional classes. She could be more than mother so long as she quite literally stays in the kitchen and allows the kitchen to extend her maternal environment.

This scene may feel far from the advances and demands being made concurrently in the 1960s Women's Liberation movement, and indeed, it may be a response to it. The threats and temptations of labor outside the home, often phrased as a panic over the morality of white middle-class women, meant that, as during World War II, the home once again had to compete with waged labor outside it. By 1970, 40 percent of American women worked outside the home, making childcare even more of a crisis and increasingly one for white mothers (one that only intensified across the remainder of the twentieth century and into our present, for Black mothers had long been at work out of the home).[52] These visions of home computing pointed a way forward by pointing a way backward: keep a (white) wife stimulated just enough, and purport to lighten her burden sufficiently, and she will continue to manage the work of both housewifery and, especially, intensive mothering. The contrary message was also evoked: now the home can shoulder mothering, too, and in fact children have two mothers (as in the Bradbury story): one of flesh and blood and one of increasingly attentive wire. For the whole family, but particularly the mother at its center, efficiency is evoked. For the children, the key word is amusement. The dream of a computerized future where a mother does not have to add traveling time to the store, and so can do more to better herself—even if just for a "diversion"—does not address itself to the childcare problem. It simply deletes it by allowing a mother to have multiple identities in the time and space in which she mothers, if she can manage it.[53]

But in the fantasy laid out by advertising for these early home computers, the housewife is set free. The problem of childcare labor is transmuted into another joy provided and justified by devotional intensive mothering. Or, domestic computing preserves gendered labor and reproduces domestic confinement while simultaneously presenting the female occupant more as wife than mother. In what can appear as a liberatory gesture, corporations furnishing homes seek to free mother from one subset of her "duties" (the word choice itself emphasizing moral compulsion and obfuscating labor) via the mechanical.

After middle-class, and especially white, women increasingly join the workforce (a long process begun in earnest during World War II, jumping again in 1970, and peaking in 1999), proto-smart homes then took on waged work exactly to return it to the domestic enclosure, building telework capacity and work-from-home solutions into their systems.[54] In the same moment that universal childcare was voted down, smart homes present themselves as an uncanny, individualistic solution—if only for the lucky few. Such uneven technologization ensured the maternal function precisely by repressing it from its imagination of a dual-career future (as mother-housewife and a PhD), aided by telecommuting and a networked home. The computerized future might allow for education and profession, but only so long as it is coupled with an explicit maintenance of the boundary of the home and therefore, implicitly, with maternal labor.

IN THE PLEASURE DOME

"Imagine a house with a brain—a house you can talk to, a house where every room adjusts automatically to match your changing moods. Imagine a house that is not just a backdrop but an active partner in your work, your family life, and your leisure activities. In short, imagine a house that is also your friend."⁵⁵ This is Xanadu, the Computerized Home of Tomorrow, which opened its doors in 1983 in Florida, strategically just a few miles from Disney World's Epcot Center. Unlike previous smart homes, which were more model and entertainment than functional domicile, Xanadu's Yale-trained architect, Roy Mason, wanted to make a walk-through instruction manual for how to make a smart home today. Mason saw this blurring of architecture and sociality as the answer to the problems of the American family: loneliness, boredom, frustration, parent-child conflict, and general depression.⁵⁶ The result proposes automation as remedy and liberation.⁵⁷

Like Bradbury's nightmare with which this book opens, Xanadu's "electric hearth" was run by a "House Brain" to control "life support functions." The House Brain is, in a way, another term for mother. Staffed by an EH (Electronic Host) dubbed Robutler (that remediated paid servant labor; see chapter 1), the house aims to take over a wide variety of tasks, but only as the owners wished. The total electric brain, not just the Robutler, was a personal assistant, presaging the integration of Alexa and Siri into smart home systems and meant to provide customizability and, crucially, effectivity in managing middle-class life.⁵⁸ The home could call local plumbers as well as neighbors, offered tele-medicine, menu planning, and even outfit selection via an automatic clothes retrieval closet. Much like Monsanto's House of the Future, Xanadu invited in thousands of guests each day to see *a* possible home of tomorrow—augmented by technology that was soon to become mainstream yet customizable to conform to the changing needs and shapes of American families.

Xanadu's promotional materials are more directly critical of systemic problems as they cross the front door and appear in the home, taking on areas of social crisis from population to urban planning, gender, and ecology. The Home for Tomorrow was just that—a material thinking-through of the differences between one generation and the next, whereas the House of the Future sought to arrest norms while updating the devices supporting and controlling them. Xanadu's materials also indicate awareness that parents were now by and large both employed full-time out of the home (some 60 percent of American families) and more egalitarian in their splitting of parenting (although it is worth noting that this has been shown to be yet another fantasy).[59] While the second baby boom had led to a housing shortage, the materials reminded prospective consumers that only 60 percent of households had children at home under the age of eighteen, and so home design must adjust away from prescriptive norms by becoming extensively customizable.[60] Xanadu asserted that a home with children's rooms was outmoded and a sign of values past, but reluctantly upheld them, while enthusiastically addressing new forms of paid labor—seeking to shrink transportation time (an answer for both sprawl and traffic as well as climate change) by building into its foundation telework capacities.

In order to substantiate a functional home of tomorrow, Xanadu's builders made use of some sleight of hand, to say the least. As media historian Laine Nooney explains, "Home computers were the first consumer technology since the television to demand a major rearrangement of domestic space; yet Xanadu negotiates this by making the computer disappear. Xanadu did not fake the technology of the times—everything here was an off-the-shelf consumer good—but it did try to use architecture to give the impression that these technologies were seamlessly integrated when they weren't."[61] This mixed temporality is the classic remit of the model home: the space is staged to allow its viewers to think of bringing these systems into their own lives *now*, while also presenting a new ideal of futuristic living.

The house pointed its way to a self-configured flexibility and customization that would become the hallmarks of smart homes in our contemporary moment, including an office that retains a domestic feel, while also gesturing backward to the mid-century notion that a woman is a manager over the automated processes in her own home.[62] If she could manage the domestic more efficiently, she could manage work from there too. Part of this putative efficiency involves taking the mother out of the car and putting her back at home, hence the feminized emphasis on tele-activities and possibilities throughout.[63] Cowan reminds us that women of this class and moment spent more time in their cars than anywhere else: "it was the vehicle through which she did her most important work, and the work locale where she could most often be found."[64] While this emphasis on telework *can* be ungendered,[65] flexible hours and commute-free work are often aimed at women in the broader culture to allow them, as Lynn Spigel notes, to multi-task across work and care, especially in order to juggle childcare duties with waged work.[66] In reality, telework is an impossibility without access to ongoing childcare, and its absence disproportionately harms women's productivity or forces them to leave the workforce altogether.

If telework is part of the total system of Xanadu, it, much like the domestic computing projects of the 1960s, extends the fantasy of "having it all" under the sign of plasticity, ease, and efficacy of networked life.[67] The children of Xanadu are assumed to be school-aged—far past demanding total attention and care. For older children, the mother's work and school schedules are understood to overlap, and they might make use of a carpool; when children are home, they will have homework to keep them occupied.[68]

That's not to say that children are not manifest in the design for Xanadu. The home does explicitly take care of its children and sees its innovations in children's spaces as planning at its "most imaginative." Pleasure, care, and control collapse again in Xanadu, whose children are no longer just off the kitchen (as in mid-century homes) but granted the status of autonomous occupants of its system, more akin to the prewar nursery. Instead of a nurse of flesh and blood accompanying the children, the home becomes caretaker. The children's room is a self-contained subsystem meant to nurture and occupy them—a version of Bradbury's playroom come to life (see figure 6.2). Using a version of NASA's windshield simulator, the children's room features a television monitor that allows children to play in any environment they might wish: under the sea, in outer space, "the wild west."[69] The beds, too, are smart, massaging children to sleep and waking them up when it is time for breakfast. Xanadu presents itself as something that "can literally grow up with a child" as its sibling, but also as "babysitter, playmate, teacher, and even a counselor to be lived with and loved."[70] For the first time, a smart home seems aware of the multiplicity of care directed toward children historically—from older sisters and grandmothers to waged and unwaged caregivers. Here, dispersed caregiving in not expressed in kin terms, and Xanadu never suggests that it is mother, for mother is in the next room. The crisis of child*care* was addressed directly via child amusement, pacification, and play, allowing a mother to work frictionlessly from home just one room away, while children were entertained nearby.

Figure 6.2
Xanadu's children's room, Xanadu pamphlet, author's personal collection.

Complete systems that follow in the lineage of the Home Electrical and the dream homes of tomorrow and the future still do not, via any technosolutionism, foretell a future where devices prioritize the remediation of motherhood. Even the 2008 Disney remake of the House of the Future, a $15 million project called the Dream Home designed by Taylor Morrison, and a decidedly self-conscious sequel to its predecessor, makes the same gesture. Like all the previous smart homes I have detailed in this partial lineage, a particular kind of family (white, middle- or upper-middle-class) is imagined as living here, as indicated by which rooms, aesthetics, and living arrangements are featured. The Dream Home takes this a step further: it has a narrative, and a fictional family living in it: the Eliases (Elias is the anglicization of Elijah—the prophet and messenger, as well as a pun on "alias"). The shift from "the future" to "dream" is in keeping with the fact that much of the technology is not speculative, succumbing to the same fate as the Home of Tomorrow; it is commonplace for some middle- and upper-middle-class families already (about which the designers said, "When we began many years ago, many of these things *were* speculative. A vast

233 *Future, Tomorrow, Dream, Smart*

majority of everything we were working on has become real"[71]). The Dream Home is, like all model homes, a message from our present and our past, not the future.

Filled with corporate sponsorship (from the likes of Microsoft and Hewlett-Packard), the Dream Home's children's rooms are in the mode of Xanadu and belong to a tween or teen. Smart technology is made apparent in every corner: smart intercoms, cameras, several touch screens, a computer, smart environmental controls, and a large screen that covers two full walls upon which customized images appear, recalling Bradbury's "nursery" once more. Like many Disney innovations, the home has its own song, set to a catchy tune: "Step through the door of your new home / Robotically cleaned from floor to dome / Your lawn's being trimmed automatically / Now there's more free time for you and me."[72] The free time for "you and me" gestures again at the liberation of women (and perhaps men, regarding key gendered tasks like yard work) from the dreary aspects of family life. The song obscures that certain unwaged labors are being mitigated to allow other unaddressed and unwaged labor to continue; mothering (or fathering) gets condensed and contracted to *free* time spent unmediated (the human-to-human relationship of "you and me") thanks to the care of the home provided by the home itself. This is the mystification of the labor of motherhood at the hands of media-as-mother.

TOIL, COST, GRIND

Across the twentieth century, the crisis of childcare has been attended by anxieties about what care and its providers might bring with them other than care itself: whether the nanny panic of the early twentieth century surrounding the introduction of "the wrong" maternal figure, which has never abated (see chapter 4),[73] or our current techno-panic about how screen media *replace* direct parenting (see chapter 5), non-maternal care is always under suspicion (see chapter 1).[74] Even as technology *does* seek to alleviate the work of mother or motherhood's disciplining functions, either via the gig economy or by devices (including a 2018 smart home hub called simply "Mother"),[75] the social and psychical crisis of childcare remains tactically, materially unimagined, reliant on the fantasies of mid-century economics and psychology and a commitment to intensive, most frequently white middle- and upper-class, forms of mothering.

As Arlie Hochschild notes in *The Second Shift*, marital roles break down into three types: traditional, transitional, and egalitarian. In brief, traditional families uphold normative divisions of labor, egalitarian households aim (but do not usually accomplish) to split care duties, and transitional families fall in between.[76] These types of gender practices are, according to Hochschild, classed, with middle and upper classes moving toward egalitarian practices (again, often in name only).[77] Single parenthood is erased as are families that share custody of children. Perhaps it is no accident that smart nursery devices are first publicized and sent to market almost exclusively by *fathers* for their own children—and often on behalf of their wives (for more on this history, see chapter 2). Across the American century, the smart home and its predecessors have sought to alleviate the housework of the traditional family while leaving mothering to the mother. A century of automating women's labor shows that early childcare is the final, untouchable bond to the home, a fact of nature at the home's center that architects do not dare redesign, even though there are many devices they could import into their plans. Augmented parenting is never displayed as part of the ideal—whether in the form of paid attention and care (a nanny, a babysitter) or electronic help. Smart homes are, after all, a highly classed and idealized form of prescriptive architecture, which is why they aggressively market themselves via the white, straight, two-parent, two-child family form. As the *ideals* of gender norms and family structures begin to shift, and the ideal includes dad doing his share, our most contemporary smart nursery items address the egalitarian household: if dad is getting up in the night too, one might think it best to automate the rocking and tending of a sleepless baby. Early motherhood and its environment can be automated, but only when they have become a distributed responsibility and shared space. When the homes of tomorrow will dream of reflecting this change remains to be seen.

For now, we are instructed, that mother remains pure because she is already a device herself, infinitely rechargeable, and always on.

CODA: ANALOG KIDS

Waldorf families. Montessori families. Reggio families. The world of alternative schooling begins at birth and merges with nuclear familial identity. Years before children are ready to enter the classroom, systematized play makes its way home to families seeking to prepare their child against the grain of contemporary schooling—and society. On Etsy, over targeted ads, and via subscription kits, parents may have these philosophies, in the form of objects for play, marketed to them and made easy—if dear—to purchase. These toys, whether a faceless soft doll in red or blue, a wooden rattle, or a mirror, are set out to soothe parental anxiety about a best beginning. Baby's play prepares the family for the school to come, and the world that opens after that. They are the first site of parental control and parental vulnerability to expertise.

Alternative education systems are not identical for being outside the mainstream and costly to purchase: some privilege notions of play-as-work (Montessori), others focus on symbolic language (Reggio). But the question of whether to exclusively purchase wooden toys or design a sparse play space is often divorced from the deeper philosophies that subtend these forms of play. Yet play is fundamentally a space of learning, and the depth of these teachings is often elided.

*

In 1919, against the backdrop of the Spanish flu, countless political uprisings on the left and right, and the aftermath of World War I, Rudolf Steiner opened the first Waldorf School in Stuttgart, Germany.[1] Steiner himself grew up adjacent to the crashing speed of modernity—as the child of a railway worker—and the educational philosophy he developed rejected contemporary technology. Instead, traditional education was preserved in the present to safeguard it for the future; students spent time out of doors to learn from nature and worked on handicrafts. Additionally, the school took its curricular focus, governance, and classroom design from Steiner's set of theories, the basis of which formed the anthroposophical movement: an esoteric occultism, replete with systematic understandings of reincarnation, spirit worlds, and higher realms.[2] Temperaments of children were understood via the typology of humors (phlegmatic, melancholic, sanguine, choleric); these were discernable, pre-ordained, and thus schematically addressed in the classroom to engender harmony. And a system of movement called "Eurythmy" played a large role, not just as physical education, but as medico-spiritual intervention to palliatively care for the kids.[3] In that first year of the school, the children likely would have practiced a specific pattern from Steiner's growing repertoire, which he taught widely to address the pandemic that raged outside: "The Immune Sequence." By coordinating sound, dance, and intention, its practitioner was thought to become their best, truest self, *and therefore* physically immune.

The notion that coordinated song and dance could provide what modern medicine did not may seem alarming to those of us who are in the midst of an ongoing pandemic, and a parallel epidemic of mis- and disinformation that keeps many from getting vaccinated (let alone medical access). But buried in Rudolf Steiner's nearly 400-volume corpus, commingled with his wide-ranging obsessive objects of study, are countless theories of viral illness and epidemiology. Steiner argued that illness was not just the result of pathogenetic spread or due to microorganisms. There was an extra-material reason for illness to present in its host: one had to already be susceptible to it via extant karma. Medical technologies and interventions—especially vaccines, which were rapidly advancing in efficacy across the last third of his life—were deemed to "make people lose an urge for a spiritual life," especially those raised outside anthroposophy (not that Steiner was keen on his followers making use of them, to say the least).[4] He called being inoculated against, say, smallpox, as being "clothed with a phantom."[5]

Steiner's theories are subtended by an admixture of quasi-Buddhist and Calvinistic thought, influenced by Germanic Romanticism and Nationalism. As with the temperaments of children, the wider outcomes of this life are already written. This extends beyond eventual destiny. Part of the anthroposophical understanding of susceptibility, too, was subtended by a karmic understanding of the hierarchy of race. Put bluntly, Steiner saw whiteness as the highest form of reincarnation; the "stagnation" of a soul was made visible by melanin. And these stagnated souls were, as Steiner had it, one way illness spread—and were to be faulted for not just their own sickness but that of society, a culture, he argued, that was characterized by attendant lack of spiritual wellness, abuses of the environment, and perhaps above all, its ruthless modernity, all of which provided a ripe environment for contagion.[6] The problem posed an endless feedback loop. Rigorous care of the individual soul was to be the cure for the body and mind, and one that could scale to a school, a nation, a world.

Behind the overt educational philosophy lives the wider meaning of techno-refusal. Although the Nazis closed nearly all Waldorf schools in the 1930s and 1940s, as historian Peter Staudenmaier shows, many of Steiner's followers—including his widow—had close ties to the Party where the occultist-karmic philosophy was taken up, both strategically and in deeper belief: both systems of thought worked to bring the world into "harmony" and achieve "regeneration."[7] Others interpret Steiner's teaching as warning of *exactly* what the Nazis had wrought—via technology. Yet much of Steiner's works and literatures were "cleaned up" to delete these mentions of race when they were translated into English, which perhaps is why many of the connections (and disconnections) between Nazism, Fascism, and this spiritual science have continued to be debated (with some in the contemporary US Waldorf movement admitting that the remarks are indeed "unfortunate").

Piloted as the school for the children of the workers at the Waldorf-Astoria Cigarette factory in Stuttgart, Germany, in 1919, the first Waldorf was the first comprehensive, coeducational space for children of all ages, classes, and abilities mixed in the same classrooms. Two thirds of the students were children of factory workers; the last third were families who followed the movement. It was, in essence, a laboratory for an alternative spiritual life and its regimes—free from modernity while absolutely born of its labors. Additional schools followed in Germany, the Netherlands, and England. Imported into the United States, in 1928 the first Rudolf Steiner school opened in New York City, and still operates today.

*

In our contemporary moment, Waldorf schools perhaps most frequently come to mainstream attention for their focus on anti-tech education, as well as for the anti-vax movement that is attracted to this wider philosophy. But this is a redundancy. Vaccines are technology. Whether in the case of a Waldorf school at the epicenter of a measles outbreak in New York City or during the COVID-19 pandemic, some Waldorf parents have rejected vaccination on behalf of their children.[8] Estimates vary, but some 60 percent of Waldorf children are unvaccinated against infectious illness for which there are routine childhood vaccines.[9] Where states grant exceptions for religious or philosophical reasons, rates are higher.

Of course, Steiner lectured on his theories on behalf of *all* his followers (and warned those who did not follow his tenets)—adults and children alike. But children could be saved before they encountered too much contagion, and moved toward their higher selves, *if* given the right conditions. For Steiner, the total system of a school could safeguard the soul and confer its own lifelong protections—the opposite of inoculation yielding the same results: a spiritual vaccine. If society maintained the wrong environment for this pursuit—and about this many would agree—Steiner's rejection of technology, and modernity's pervasiveness, means that he is frequently read as *anti*-modern. Perhaps, instead, this is the other side of modernity. To buy into these teachings is to locate a rationale for alternative possibilities within what techno-culture has wrought. That method is sent home with children in the form of rules for families—as well as consumer goods, however nostalgic they appear for a time gone by, disciplining the family that in turn participates in the life of the school. For Waldorf, parents safeguard the teachings at home, while the teacher issues them.

*

Mainstream parents, too, have been constantly troubled about how much media is too much—but also how much is enough. As we have seen across this book, pediatric panics about both the impact of media technologies *and* their content have abounded across the last century, from worrying about optical games, "bug eyes" related to television consumption, violence and videogames, or the destruction of empathy on social media.

Steiner's adherents, whether they are "Waldorf families" or in the anthroposophical movement or both, support a form of collective individuality, which, as media historian Laura Portwood-Stacer argues, is frequently at play in the rejection of media and technology. Grounded in the notion of the highest form of spirit, and of pre-destiny, they can reject "herd mentality" and in doing so compromise herd immunity, even as leaders within the movement call for the widespread acceptance of vaccines to stem the tide of COVID-19.[10] As science and technology studies scholar Morgan Ames writes on the twinned rejection of techno-education and vaccines by wealthy parents in the twenty-first century, "Private schools almost by definition have to craft stories that appeal to privileged strivers anxious about their children's futures. Some of these stories recount how their graduates' creative brilliance was spawned in their school's tech-free environment. Related ones ply anti-contamination themes, and fetishize the purity of childhood."[11] These stories communicate technology as a threat to that spirit and individuality, and thus Waldorf schools offer a reprieve from the constant speed of society and, its subscribers argue, from the direct influence of society. Parables of contagion and techno-refusal also reprise the purity ideal that lobbies for unmediated mothering.

Steiner's philosophy, however much or little subscribed to in the contemporary, expresses a cultural reconsideration of technology's role in our twenty-first-century relationships, even if understood by its critics to be extreme. This rejection of the present is articulated on the grounds of the future: children. But here, the anti-tech purity is inextricable from a notion of blood purity. As anthropologist Danya Glabau writes, "As a metaphor, purity easily translates from necessary practices to exclusionary principles."[12]

Yet in Steiner's theories purity is spiritual, but not metaphorical. Returning to Ames's notion of the charismatic story of the purity of childhood, what does it mean if that purity is subtended by an entire system predicated on hierarchies of souls—and hierarchies of race?

This has been implicit in all Waldorf schools from the very first to those operant in the present. In 1920 Waldorf schools were meant to redress high modernism. In the 1970s, Waldorf schools were sought out by parents who might otherwise, as one *New York Times* article put it, take up "transcendental meditation and primal screaming and astrology and est and all the other byways of the human potential movement."[13] Television was frequently referred to as the corruptive force, the major target to be avoided at all costs, to the point that when one Waldorf student had a TV set wheeled into his hospital recovery room, he screamed.

*

COVID-19 posed another, quieter problem for Waldorf schools and other alternative systems that reject technology: What to do with mandated social distancing and remote education? Some Waldorf schools proceeded to move online, especially for older children; others gathered in person. Some met not at all, assuming parents would pick up the labor of educating their children. In at least one Waldorf school, the middle schoolers and high schoolers met regularly to practice "Eurythmy," with a focus on the same pattern of movement that was done during the Spanish flu epidemic of the last century, "The Immune Sequence."[14]

Some have prayers, some have panics, some have sequences. All try to keep at bay the sense that what we are putting into children has gone horribly wrong, even if that bad turn has long existed. All contain, at their core, a politics of material and maternal purity, one that selectively identifies an "outside" to its striving (or as John Milton had it, "To the Pure, all Things are Pure"). In the most recent pandemic, as Samuel Weber has recently argued, we have begun to re-see the preexisting conditions of contemporary life.[15] To those looking to ensure a better life for their children, the wrong modernity has been realized, and once again the millennialism and apocalypticism that run in Steiner's theories seems more appreciably accurate, even to nonbelievers (just look to the climate crisis). And once again, the link between Fascism and occultism—and its shared preoccupation with the infiltration of modernity—may be reprised, especially on the grounds of anti-COVID-19 vaccination theories that offer a literal form of this infiltration (whether via notions of chipping, magnetism, or supposed ruination of biological reproduction).

*

Steiner's notion that the world, and specific mispreparations of children (in his case, anyone raised outside his spiritual system), makes us ill has its parallels in other moments adamantly against the inclusion of technology in child-rearing, and the children they raise. Where Waldorf exists as an alternative, parents who otherwise might be termed progressive have sought it out. These social and parenting movements take so seriously the threat of technology as interloper that they do away with it altogether, seeking to restrict children to liberate them into a childhood understood to be pure by way of its dismediation. They reject society as it is to save society as it could be. Across the twentieth century, as life was understood to be increasingly media-driven and intershot with technology, this logic has been pervasive if less extreme than its form in Steiner's theories and institutions. If techno-medicine was the outside coming in (especially, eventually, in the case of vaccines), many of the protocols of Waldorf were about the inside coming out and further into focus. But the ideology of purity can cut both ways; the alternative does not necessarily bend toward Fascism or toward even apoliticality. Techno-refusal, especially in child-rearing, looks quite different in the hands of leftist conceptions or other utopian projects. Anti-technological childcare principles can be found in as diverse contexts as the Black radical 1970s–1980s MOVE organization, which argued that media spread the ills of white supremacy; in the back-to-the-land movement of the 1960s; in other forms of communal care that focused on raising children in nature; or in separatist movements of multiple political persuasions.[16] Many groups see media as interfering with the direct messages of their internal protocols, and often of the child's most prominent ideological communicator and first environment: their mother.

Acknowledgments

That none of us makes it alone is the subject of this book, and of course, the truth of its writing.

I have immense gratitude to my friends without whom (I wrote home) I would be bereft.

I am indebted to many colleagues for their direct support of this particular book's thinking or for your aid in shifting my conditions in writing it: Nicole Starosielski, Ella Klik, Tamara Kneese, Beth Semel, Finn Brunton, Kevin Duong, Olivia Banner, John Durham Peters, Jeff Mathias, Kelli Moore, Susan Murray, Anthony Ryan Hatch, Ilana Gershon, Nathan Ensmenger, David Henkin, Mar Hicks, Ramsey McGlazer, Gabriel Winant, Jaipreet Virdi, Sharonna Pearl, Morgan Ames, Rebekah Sheldon, Sandra Eder, Jacob Gaboury, Dora Zhang, Massimo Mazzotti, Gail De Kosnick, Marissa Mika, Cathryn Carson, Poulomi Saha, Aarti Sethi, Ussama Makdisi, Eric Falci, and Lara Wolfe. I also sat in a room at the Kinsey Institute in Bloomington, Indiana, most Friday mornings with Sarah Knott as I finished the first complete draft of this book—thank you for your friendship and for reading the manuscript at a crucial stage, and to Jeff Nagy, Fred Turner, Alex Colston, Nico Baumbach, and Bernie Geoghegan for reading the book as I revised it. I've been grateful—and very lucky—to have the continuous mentorship of Ben Kafka, Mara Mills, Elizabeth Wilson, Jeremy A. Greene, and Fred Turner.

Key funding for this book was provided by the Society of the History of Technology, which awarded the book the 2021 Brooke Hindle Fellowship.

The writing of it would have been much more precarious if not for the funding attached to that award—I wrote the lion's share of this book while contingent faculty and had no support for its research or writing otherwise. My sincere gratitude goes to the Society for their support of this work, and to the Hindle family. I am also grateful to the Abigail Hodges Publication Fund from the Dean's Office at UC Berkeley which helped me finish it.

This project simply would not exist without the archivists at a dozen or so repositories who helped me access materials during a global pandemic, often remotely. Thanks so much to Lizette Royer Barton at the Nicholas and Dorothy Cummings Center for the History of Psychology at the University of Akron, to the whole team at Researcher Services at the New York State Archives, to Deena Gorland at the American Psychiatric Association Foundation Library and Archives, and to those at the Countway Library at Harvard (with gratitude to Kaitlin Smith, who helped me access some of the Skinner material), Cornell University's HEARTH (Home Economics Archive: Research, Tradition and History), the Library of Congress Manuscript Division, the Noguchi Archives, the Victoria and Albert Museum, the New York Public Library's Schomburg Center, and Brooklyn College Archives. I am endlessly indebted to the Psychoanalytic Electronic Publishing website (PEPWEB) for making available a huge swath of the published primary sources on which this book, and all my work, relies.

I am particularly grateful to the participants and structures of the three working groups I participated in during the writing of this book: the Feminist Research Institute at UC Davis in 2020–2021, the Motherhood and Technology working group at the Columbia Center for the Study of Social Difference, and the MEDS working group at UC Berkeley. In my last year at work on the book, I was engaged in founding *Parapraxis* and the Psychosocial Foundation. Our collective work to elucidate "The Family Problem" in issue 1, and in its parallel project, Seminar 1, was crucial to the theoretical engagement with what exactly it might be that technologies are doing in the reproductive sphere both psychically and materially. I returned copy edits on the book as we finished our work on issue 04, Security—which also inflects my thinking, especially in chapters 1 and 4. My thanks again—especially to our collective—and to all the presenters, facilitators, participants, and writers who make *Parapraxis* and The Foundation what they are.

Parts of this project have been presented in various iterations at events at the Center for Social Difference and the Heyman Center at Columbia University; Indiana University's Department of American Studies; the University of Santa Cruz' Cultural Studies Colloquium; Notre Dame's Department of the History and Philosophy of Science; Johns Hopkins's Department of the History of Medicine; Kings College London; Colorado College; Yale's History of Science and Medicine Program; the Yale Law School; Light Industry; the Toronto Public Library; the University of California, Berkeley; the Psychosocial Foundation; the Special Interest Group for Computing, Information, and Society; the Society for Literature, Science, and the Arts; the Association of Internet Researchers; the Society for the Social Study of Science; the American Studies Association; and the Society for Cinema and Media Studies—many thanks to the organizers of these events and to the audiences there.

The first person to take a chance on this book, and really on me as a writer rather than just a thinker, was Jim Rutman, who is my favorite non-Knicks fan to talk basketball with, among many other things; thank you, Jim. I am grateful to continue to work with the MIT Press, which has been such an immense source of support over the last years. My special thanks to my editor there, Gita Devi Manaktala. Thank you to everyone at the press who had a hand in the making of this book: Deborah Cantor-Adams, Stephanie Sakson, Yasuyo Iguchi, Sheena Meng, Suraiya Jetha, David Olsen, and those still unknown to me. Many thanks to freelance indexer David Luljak for his fine work. To the anonymous reviewers who read the earliest and the final versions of the book—thank you for your attention to the manuscript, and to Amy O'Hearn for helping with the images and their permissions. Excerpts from this book appeared in *differences: A Journal of Feminist Cultural Studies*, *Real Life*, *Parapraxis*, and *Logic(s)*. I thank my editors there, including the entire editorial team of *differences* as well as the anonymous reviewers there. I am particularly grateful to Denise Davis, Alexandra Molotkow, and Sarah Burke, and Alex Colston.

My deepest thanks to my family and my family of affinity, especially to Geoffrey G. O'Brien, my first and last reader. And Geoffrey: thank you for making life joyous when life, let alone joy, seems an impossibility.

This book is dedicated to my mother, Lynne Zeavin, my medium.

The book is also dedicated to my son, Malachai, who calls me his habitat.

Notes

INTRODUCTION

1. Ray Bradbury, "The Veldt," in *The Illustrated Man* (Garden City, NY: Doubleday, 1951), 9.
2. Bradbury, "The Veldt."
3. Bradbury uses the term "nursery" throughout—a telling misnomer even in his moment. The "nursery" is a loose term, and absolutely dependent on historical context. In the eighteenth century, it was any room or series of rooms for a baby's or child's particular use, including, if the family was in a position to do so economically, any attendants employed to care for that child (nurses, nannies, governesses, maids, and others). The nurse or nanny would also sleep in this suite of rooms and was in charge of the overnight care (as well as care during the day) of their charge(s). By Bradbury's moment, one might have a nursery *and* a playroom, or just children's bedrooms. For more on the eighteenth century, see Karin Calvert, *Children in the House: The Material Culture of Early Childhood, 1600–1900* (Boston: Northeastern University Press, 1994); for more on the twentieth century, see Amy Ogata, *Designing the Creative Child: Playthings and Places in Midcentury America* (Minneapolis: University of Minnesota Press, 2013).
4. Bradbury is forecasting that a nursery so in sync with the fantasies—the desires—of children will prevent them from learning to tolerate frustration and separation, that technological attunement to their every whim will misprepare them for this harsh world. The Hadleys acquired the nursery for the children as a kind of cure for some unnamed form of neurosis; the nursery's technology was initially developed as a psychological diagnostic tool because it would reveal a child's innermost fantasies—a virtual reality Rorschach test packaged as both babysitter and consumer toy. This too is borne out in reality and goes both ways—consumer toys become psychiatric-grade diagnostic tools, and psychiatric tools become utilized in the mainstream as leisure objects.
5. Raymond Williams, *Television: Technology and Cultural Form*, 3rd ed. (New York: Routledge, 2003).
6. Coontz points us to "how very exceptional in the descriptive sense of the word is American Mothering in the twentieth and twenty-first century." Stephanie Coontz, "Historical

Perspectives on Family Studies," *Journal of Marriage and Family* 62, no. 2 (May 2000): 284. See also Sheila Kitzinger, *Women as Mothers: How They See Themselves in Different Cultures* (New York: Random House, 1978); Sarah Blaffer Hrdy, *Mother Nature: A History of Mothers, Infants, and Natural Selection* (New York: Pantheon, 1999) and *Mothers and Others: The Evolutionary Origins of Mutual Understanding* (Cambridge, MA: Harvard University Press, 2011).

7. Stephanie Coontz, "Historical Perspectives on Family Studies," *Journal of Marriage and Family* 62, no. 2 (May 2000): 284.

8. John Demos, *A Little Commonwealth: Family Life in Plymouth Colony*, 2nd ed. (New York: Oxford Academic, 1999; online ed. October 2011).

9. Jeffrey Sconce shows how media, mediation, and the medium (as in the clairvoyant) attached to femininity in the late nineteenth century. In this way, the first feminized mediums were telepaths, not mothers. See *Haunted Media: Electronic Presence from Telegraphy to Television* (Durham, NC: Duke University Press, 2000).

10. Mary Fissell, "Popular Medical Books," in *Oxford History of Popular Print Culture, vol 1: Beginnings to 1660*, ed. Joad Raymond (Oxford: Oxford University Press, 2011), 418–431.

11. Stephanie E. Jones-Rogers, *They Were Her Property* (New Haven, CT: Yale University Press, 2019), see chapter 5.

12. Diana Paton, "The Driveress and the Nurse: Childcare, Working Children and Other Work under Caribbean Slavery," *Past & Present* 246, no. 15 (December 2020): 27–53, https://doi.org/10.1093/pastj/gtaa033.

13. Namwali Serpell, *Stranger Faces* (Oakland, CA: Transit Books, 2020).

14. See Jules Gil-Peterson's *Histories of the Transgender Child* (Minneapolis: University of Minnesota Press, 2018), chapter 1.

15. Rima D. Apple, "Constructing Mothers: Scientific Motherhood in the Nineteenth and Twentieth Centuries," *Social History of Medicine* 8, no. 2 (1995): 161–178.

16. Molly Ladd-Taylor and Lauri Umansky, eds., *"Bad" Mothers: The Politics of Blame in Twentieth-Century America* (New York: New York University Press, 1998).

17. Carola Ossmer, "Normal Development: The Photographic Dome and the Children of the Yale Psycho-Clinic," *Isis* 111, no. 3 (2020): 515–541.

18. See Shaul Bar-Haim's *The Maternalists: Psychoanalysis, Motherhood, and the British Welfare State* (Philadelphia: University of Pennsylvania Press, 2009).

19. Thomas Freeman, "James Clark Moloney: 'Mother, God and Superego,'" *International Journal of Psychoanalysis* 36 (1955): 222.

20. This paradigm picked up speed during and after World War II, but was long in evidence before and after. The frequency of mentions of new terms and coinages centers on 1939–1970. Most often, the term "maternal environment," which will be popularized by Winnicott, is used before the mid-century diffusion and explosion of terms. In its earliest usages,

the environment is precisely *not* the mother-baby combination. The environment is, well, everything that lies outside it.

But just three years later, Edward Glover, in England, is using "Maternal environment" as the basis for a psychoanalytic characterology—based on observation and, on his own self-understanding, therefore empiric. Edward Glover, "Notes on Oral Character Formation," *International Journal of Psychoanalysis* 6 (1925): 131–154. In 1929, Ferenczi is using the phrase "*from the warm Maternal environment*" in the way that presages its more famous usage in Winnicott. The question for Ferenczi is what to make of infantile death instincts; the paper is deeply in conversation with Freud's new theorization of the death drive. Sándor Ferenczi, "The Unwelcome Child and His Death-Instinct," *International Journal of Psychoanalysis* 10 (1929): 125–129.

21. Georges Canguilhem, "The Living and Its Milieu," in *Knowledge of Life* (New York: Fordham University Press, 2008), 98–120.

22. Christina Wessely and Nathan Stobaugh, "Watery Milieus: Marine Biology, Aquariums, and the Limits of Ecological Knowledge circa 1900," *Grey Room* 75 (2019): 36–59.

23. The question of whether or not matrixes are media themselves has been written about extensively by John Durham Peters.

24. I write about this at length in "Therapy with a Human Face," *Dissent Magazine* (Winter 2022), and in "No Touching: Boundary Violation and Psychoanalytic Solidarity," *differences: A Journal of Feminist Cultural Studies* 33, nos. 2–3 (July and December 2023).

25. See the work of Martin Summers, *Madness in the City of Magnificent Intentions: A History of Race and Mental Illness in the Nation's Capital* (New York: Oxford University Press, 2019); see also Jonathan Metzl, *The Protest Psychosis* (Boston: Beacon Press, 2009).

26. See Jonathan Metzl, *Prozac on the Couch* (Durham, NC: Duke University Press, 2003).

27. As quoted in Ellen Herman, *The Romance of American Psychology: Political Culture in the Age of Experts* (Berkeley, California: UC Press, 1995), 278–279.

28. Sarah Knott, *Care and Capitalism* (Durham, NC: Duke University Press, 2025).

29. "Matrix" for Winnicott becomes a synonym for the relationship between mother and child. The environment is *the mother* full stop. See works by D. W. Winnicott, "Ego Integration in Child Development" (1962), in *The Collected Works of D. W. Winnicott*, ed. Lesley Caldwell and Helen Taylor Robinson (New York: Oxford University Press, 2016) and *The Maturational Processes and the Facilitating Environment: Studies in the Theory of Emotional Development* (Madison, CT: International Universities Press, 1965); and by Marion Milner, *The Hands of the Living God: An Account of a Psycho-Analytic Treatment* (New York: Routledge, 2010).

30. See Susanna Schmidt's excellent *Midlife Crisis: The Feminist Origins of a Chauvinist Cliché* (Chicago: University of Chicago Press, 2021) and Knott, *Care and Capitalism*, for more on the feminist elaborations of Eriksonian critique. For the debate between psychoanalytic and behavioristic accounts of mother-infant relating, and a range of responses to their theoretical differences, see the argument between André Green and Daniel Stern in *Clinical and*

Observational Psychoanalytic Research: Roots of a Controversy—Andre Green and Daniel Stern, ed. Joseph Sandler, Anne-Marie Sandler, and Rosemary Davies (London: Routledge, 2000); see also Mary Jacobus, *The Poetics of Psychoanalysis: In the Wake of Klein* (Oxford: Oxford University Press, 2005).

31. T. Berry Brazelton, Barbara Koslowski, and Mary Main, "The Origins of Reciprocity: The Early Mother-Infant Interaction," in *The Effect of the Infant on Its Caregiver*, ed. M. Lewis and L. A. Rosenblum (New York: Wiley, 1974), 49–76; Felix E. Rietmann, "Mother-Blaming Revisited: Gender, Cinematography, and Infant Research in the Heyday of Psychoanalysis," *History of the Human Sciences* 37, no. 2 (2023), https://doi.org/10.1177/09526951231187556.

32. Angelica Clayton, "Revisiting *The Body Keeps the Score*," Psychosocial Foundation, January 23, 2023.

33. The slip from media to medium, on the grounds that a mother provides mediation, is central to the construction of these theories of all-in, all-the-time, total mothering. As Guillory suggests, mediation is an "abstract process"—the move from "'agent' to 'agency'" an "impersonal process." John Guillory, "Genesis of the Medium Concept," *Critical Inquiry* 36 (Winter 2010): 324.}

34. Guillory, "Genesis of the Medium Concept."

35. See Meredith Bak, *Playful Visions: Optical Toys and the Emergence of Children's Media Culture* (Cambridge, MA: MIT Press, 2020).

36. Dylan Mulvin, *Proxies: The Cultural Work of Standing In* (Cambridge, MA: MIT Press, 2021).

37. Guillory, "Genesis of the Medium Concept."

38. Marie Briehl, "Review of *Parents' Questions: By Staff Members of the Child Study Association of America* (New York: Harper Bros., 1936)," *Psychoanalytic Quarterly* 6 (1937): 371–375.

39. For more on this work, see David E. Morrison, "Kultur and Culture: The Case of Theodor W. Adorno and Paul F. Lazarsfeld," *Social Research* 45, no. 2 (1978): 331–355; Josh Shepperd, "Theodor Adorno, Paul Lazarsfeld, and the Public Interest Mandate of Early Communications Research, 1935–1941," *Communication Theory* 32, no. 1 (February 2022): 142–160; Thomas Y. Levin and Michael von der Linn, "Elements of a Radio Theory: Adorno and the Princeton Radio Research Project," *Musical Quarterly* 78, no. 2 (1994): 316–324.

40. Arthur T. Jersild, Frances V. Markey, and Catherine Jersild, *Children's Fears, Dreams, Wishes, Daydreams, Likes, Dislikes, Pleasant and Unpleasant Memories: A Study by the Interview Method of 400 Children Aged 5 to 12* (New York: Bureau of Publications, Teacher's College, Columbia University, 1933).

41. Margaret Schoenberger Mahler, "Tics and Impulsions in Children: A Study of Motility," *Psychoanalytic Quarterly* 13 (1944): 430–444.

42. Psychoanalysts, specifically, were belatedly interested in television. In A Report of an Open Forum held at the 29th International Psycho-Analytical Congress, London, July 1975, the chairman of the forum was Robert S. Wallerstein, San Francisco; and the forum members

were Clemens de Boor, Frankfurt; Luis Feder, Mexico; Edward Joseph, New York; and Thomas Main, London. Television was stressed as something that as yet psychoanalysts had not taken into account, but which was certain to have an impact upon child development, because of the greater ease of auditory and visual communication via television.

43. Tom Engelhardt, "The Primal Screen," *Mother Jones*, May/June 1991.

44. Neil Postman, "Media Ecology Education," *Explorations in Media Ecology* 5, no. 1 (March 2006): 5–14.

45. Lynn Spigel, *Make Room for TV: Television and the Family Ideal in Postwar America* (Chicago: University of Chicago Press, 1992), 36.

46. Erving Goffman, *Asylums: Essays on the Social Situation of Mental Patients and Other Inmates* (New York: Doubleday and Anchor Books, 1961).

47. Goffman, *Asylums*.

48. Elaine Tyler May, *Homeward Bound: American Families in the Cold War Era* (New York: Basic Books, 1988).

49. Sophie Lewis, *Full Surrogacy Now: Feminism against Family* (London: Verso, 2021).

50. Lisa Baraitser, *Maternal Encounters: The Ethics of Interruption* (London: Routledge, 2009).

51. Hortense J. Spillers, "Mama's Baby, Papa's Maybe: An American Grammar Book," *Diacritics* 17, no. 2 (1987): 64–81; Alexis Pauline Gumbs, China Martens, and Mai'a Williams, eds., *Revolutionary Mothering: Love on the Front Lines* (Oakland, CA: PM Press, 2016).

52. Sarah Knott, *Mother Is a Verb* (New York: Farrar, Straus and Giroux, 2019).

53. See the work of Christina Sharpe, such as *In the Wake: On Blackness and Being* (Durham, NC: Duke University Press, 2016), and Saidiya Hartman, *Lose Your Mother* (New York: Farrar, Straus and Giroux, 2008).

54. See Dorothy Roberts's work on family policing and its abolition across the last thirty years, most recently, *Torn Apart: How the Child Welfare System Destroys Black Families and How Abolition Can Build a Safer World* (New York: Basic Books, 2023), and Michal Raz, *Abusive Policies: How the American Child Welfare System Lost Its Way* (Chapel Hill: University of North Carolina Press, 2020).

55. Angela Y. Davis, *Women, Race, and Class* (London: Women's Press, 1982), chapter 17.

56. Lynn Spigel, "Designing the Smart House: Posthuman Domesticity and Conspicuous Production," *European Journal of Cultural Studies* 8, no. 4 (November 1, 2005): 419–420.

57. What the role of the American mother was understood to be in 1910 or so depended on many factors: race, class, geography, immigration status, disability, and ability. Some conditions of life were essentially universal: for example, legally, women didn't have rights to their own wages. Only some 4 percent of married women worked outside the home.

58. Tanya Klich, "The New Mom Economy: Meet the Startups Disrupting the $46 Billion Millennial Parenting Market," *Forbes*, May 10, 2019.

59. Sharon Hays, *The Cultural Contradictions of Motherhood*, (New Haven, CT: Yale University Press, 1996).

60. Fredric Jameson, "The End of Temporality" *Critical Inquiry* 29, no. 4 (Summer 2003): 695–718; Klaus Theweleit, *Object-Choice (All You Need Is Love)* (London: Verso, 1994).

61. Friedrich Kittler, *Dichter/Mutter/Kind* (Munich: Wilhelm Fink Verlag, 1991). Or, as Bernard Geoghegan glosses, "An application [of the technical a priori] so radical it proceeds without the incidental mediation of circuit boards, telephone relays, or human programmers, finding instead the laws of communication technics at work in biology, romantic intimacy, and child-rearing?" Private communication, February 9, 2024.

62. Phillip Maciak, *Avidly Reads: Screentime* (New York: New York University Press, 2022).

63. Melinda Cooper, *Family Values: Between Neoliberalism and the New Social Conservatism* (New York: Zone Books, 2017), 7.

CHAPTER 1

1. "Lindbergh Baby Kidnapped from Home of Parents on Farm Near Princeton; Taken from His Crib; Wide Search On," *New York Times,* March 2, 1932.

2. "Lindbergh Baby Kidnapped from Home."

3. "Lindbergh Baby Kidnapped from Home"; Paula S. Fass, *Kidnapped: Child Abduction in America* (Oxford: Oxford University Press, 1991), 7, 88–99.

4. "Betty Gow Takes the Stand," *Manchester Evening Herald,* January 2, 1935.

5. Newman F. Baker, "The Case of Violet Sharpe," *Journal of Criminal Law and Criminology (1931–1951)* 23, no. 2 (1932): 166–168.

6. Baker, "The Case of Violet Sharpe."

7. Richard T. Cahill Jr., *Hauptmann's Ladder* (Kent, OH: Kent State University Press, 2014), 16.

8. "Charles Lindbergh's Baby Found Dead beside the Princeton-Hopewell Road," *Daily News,* May 13, 1932.

9. "Charles Lindbergh's Baby Found Dead."

10. "Lindbergh Baby Kidnapped from Home of Parents on Farm Near Princeton."

11. Thomas W. Trenchard and Mike Stewart, eds., *The State vs. Hauptmann,* vol. 1: *The Lindbergh Baby Kidnapping and Murder Trial Transcripts.*

12. Trenchard and Stewart, *The State vs. Hauptmann.*

13. "Hauptmann Put to Death for Killing Lindbergh Baby Remains Silent to the End," *New York Times,* April 4, 1936.

14. Brian Connolly, "The Nuclear Family: How a Complex Became a Norm," *Parapraxis* 1 (December 2022).

15. What constituted family change, and how it changed, was the focus of intense historiographical and demographic debate in the 1990s especially.

16. Neda Atanasoski and Kalindi Vora, "Surrogate Humanity: Posthuman Networks and the (Racialized) Obsolescence of Labor," *Catalyst: Feminism, Theory, Technoscience* 1, no. 1 (2015): 4.

17. Atanasoski and Vora, "Surrogate Humanity."

18. See Shellee Colen, "'With Respect and Feelings': Voices of West Indian Child Care Workers in New York City," in *All American Women: Lines That Divide, Ties That Bind*, ed. Johnnetta B. Cole (New York: Free Press, 1986), 46–70; and Rayna Rapp and Faye Ginsburg, *Conceiving the New World Order: The Global Politics of Reproduction* (Berkeley: University of California Press, 1996). See Grace Chang's *Disposable Domestics: Immigrant Workers in the Global Economy* (Chicago: Haymarket Books, 2016).

19. Feminism too has had a long, complicated relationship to technologies that describe a different future. As Judy Wajcman wrote in her landmark book *TechnoFeminism*, feminists are "torn between utopian and dystopian visions of what the future may hold. In both these scenarios, the future is crowded with automata, androids, and robots.... What might these imaginings about the future reveal about contemporary gender relations?" Wajcman, *Techno Feminism* (Cambridge: Polity, 2004), 3. This chapter does not take up the feminist or utopian or radical experiments with collective living, or architectural practices for the same, that address childcare centrally. As Dolores Hayden writes, in this same postwar period, there were European attempts to add human-provided childcare to housing specifically so that women who had joined the workforce during the war might maintain their jobs. This was attempted via collective living elements and daycare centers in apartment buildings—and these ideas were reported on in the United States but not adopted. Dolores Hayden, *Redesigning the American Dream: The Future of Housing, Work, and Family Life* (New York: Norton, 2002), 111–113. See also Shaowen Bardzell, "Utopias of Participation: Feminism, Design, and the Futures," *ACM Transactions on Computer-Human Interaction* 25, no. 1 (February 2018), https://doi.org/10.1145/3127359.

20. Sonya Michel, "The History of Child Care in the U.S.," Social Welfare History Project, Virginia Commonwealth University, January 19, 2011, https://socialwelfare.library.vcu.edu/programs/child-care-the-american-history.

21. Michel, "The History of Child Care in the U.S."

22. Sarah Knott, "Theorizing and Historicizing Mothering's Many Labours." *Past & Present* 246, no. 15 (December 2020): 1–24.

23. See Theda Skocpol, *Protecting Soldiers and Mothers: The Political Origins of Social Policy in the United States* (Cambridge, MA: Harvard University Press, 1992). For postwar implications, see Emilie Stolzfus, *Citizen, Mother, Worker: Debating Public Responsibility for Child Care after the Second World War* (Chapel Hill: University of North Carolina Press, 2003).

24. Anne Boyer, *The Undying: A Meditation on Modern Illness* (New York: Farrar, Straus and Giroux, 2019).

25. Mirroring the schism in which labor histories and histories of mothering are often an either/or proposition, in which the nanny as mother is seemingly intolerable, the nanny's role in child development is undermined to preserve the fantasy that loving care remains nonfungible, as if to do the work of standing in for someone is not to become them, nor to complicate their return.

26. See Mrs. John Lane: "The Servant Question in England," *Harper's Bazaar* 39 (1905): 115–119; Peggy Wood: "A House in London," *Saturday Evening Post*, March 7, 1931, 109–113; and Lucy Maynard Salmon, *Domestic Service in the United States* (London: Macmillan, 1901); "Nurses for Children," in *Englishwomen's Year Book and Directory*, ed. Geraldine Edith Mitton (London: A. & C. Black, 1910).

27. Mary L. Read, "The Professional Training of Children's Nurses," *Teachers College Record* 11, no. 7 (1910), https://journals.sagepub.com/doi/10.1177/016146811001100702.

28. Lynn Spigel, "Designing the Smart House: Posthuman Domesticity and Conspicuous Production," in *Electronic Elsewheres: Media, Technology, and the Experience of Social Space*, ed. Chris Berry, Soyoung Kim, and Lynn Spigel (Minneapolis: University of Minnesota Press, 2009), 55–92.

29. Martha Haygood Hall, "The Nursemaid: A Social-Psychological Study of the Group," PhD diss., University of Chicago, 1931, 21.

30. Stephanie E. Jones-Rogers, *They Were Her Property: White Women as Slave Owners in the American South* (New Haven, CT: Yale University Press, 2019), 102.

31. See also Gal Ventura, *Maternal Breast-Feeding and Its Substitutes in Nineteenth-Century French Art* (Leiden: Brill), especially "From Sanctity to Promiscuity: The Wet Nurse"; and Emily West and R. J. Knight, "Mothers' Milk: Slavery, Wet-Nursing, and Black and White Women in the Antebellum South," *Journal of Southern History* 83, no. 1 (2017): 37–68.

32. Jones-Rogers, *They Were Her Property*, 103, see footnote 10.

33. See Jonathan Gathorne-Hardy, *The Unnatural History of the Nanny* (New York: Dial Press, 1973), 42–43.

34. Karin Calvert, *Children in the House: The Material Culture of Early Childhood, 1600–1900* (Boston: Northeastern University Press, 1994), 122–123.

35. See Jessica Martucci, *Back to the Breast: Natural Motherhood and Breastfeeding in America* (Chicago: University of Chicago Press, 2015).

36. Christina Hardyment, *Dream Babies: Childcare Advice from Locke to Spock* (London: Jonathan Cape, 1983), 90.

37. Lynn Y. Weiner, *From Working Girl to Working Mother: The Female Labor Force in the United States, 1820–1980* (Chapel Hill: University of North Carolina Press, 1985), 31.

38. Catharine E. Beecher and Harriet Beecher Stowe in *The American Woman's Home* (New York: J. B. Ford, 1869). For more, see Kathryn Kish Sklar, *Catharine Beecher: A Study in American Domesticity* (New York: W. W. Norton, 1976).

39. Beecher and Stowe, *The American Woman's Home*.

40. Calvert, *Children in the House,* 125.

41. See Celestina Wroth, "'To Root the Old Woman Out of Our Minds': Women Educationists and Plebeian Culture in Late-Eighteenth-Century Britain," *Eighteenth-Century Life* 30, no. 2 (April 2006): 48–73.

42. For a rich history of women's work outside the house, see Weiner, *From Working Girl to Working Mother*, especially chapter 2, "The Discovery of Future Motherhood." For the long history of white women's subjugation of Black women who care for their children, see Jones-Rogers, *They Were Her Property*, 110–120. For more on the intersection of reproductive labor by enslaved women with gender and race, see Jennifer L. Morgan, *Laboring Women: Reproduction and Gender in New World Slavery* (Philadelphia: University of Pennsylvania Press, 2004), especially chapter 4.

43. Calvert, *Children in the House*, 124.

44. Viviana Zelizer, *Pricing the Priceless Child* (Princeton, NJ: Princeton University Press, 1985), 10.

45. Robin Bernstein, *Racial Innocence: Performing American Childhood from Slavery to Civil Rights* (New York: New York University Press, 2011), 36.

46. Bernstein, *Racial Innocence*.

47. Mary Pat Brady, *Scales of Captivity: Racial Capitalism and the Latinx Child* (Durham, NC: Duke University Press, 2022), 7; Zakiyyah Iman Jackson, *Becoming Human: Matter and Meaning in an Anti-Black World* (New York: New York University Press, 2020).

48. Zelizer, *Pricing the Priceless Child*.

49. M. E. O'Brien, "To Abolish the Family: The Working-Class Family and Gender Liberation in Capitalist Development," *Endnotes* 5 (2019).

50. Zelizer, *Pricing the Priceless Child*, 6.

51. Hall, "The Nursemaid," 21.

52. Daria Colombo, "'Worthless Female Material': Nursemaids and Governesses in Freud's Cases," *Journal of the American Psychoanalytic Association* 58, no. 5 (2010): 835–859.

53. Carolyn Steedman, *Labors Lost: Domestic Service and the Making of Modern England* (Cambridge: Cambridge University Press, 2009), 320.

54. Sigmund Freud, "The Psychotherapy of Hysteria from Studies on Hysteria," in *The Standard Edition of the Complete Psychological Works of Sigmund Freud*, ed. J. Strachey (London: Hogarth Press, 1893), II: 274.

55. Sigmund Freud, "The Aetiology of Hysteria," in *The Standard Edition of the Complete Psychological Works of Sigmund Freud*, ed. J. Strachey (London: Hogarth Press, 1896), III: 208; Sigmund Freud, "Heredity and the Aetiology of the Neuroses," in *The Standard Edition of the Complete Psychological Works of Sigmund Freud*, ed. J. Strachey (London: Hogarth Press, 1896), III: 152.

56. Lucy Delap, *Knowing Their Place: Domestic Service in Twentieth-Century Britain* (Oxford: Oxford University Press, 2011).

57. Ruth Schwartz Cowan, *More Work for Mother: The Ironies of Household Technology from the Open Hearth to the Microwave* (New York: Basic Books, 1985), 36.

58. Russell A. Powell, Nancy Digdon, Ben Harris, and Christopher Smithson, "Correcting the Record on Watson, Rayner, and Little Albert: Albert Barger as 'Psychology's Lost Boy,'" *American Psychologist* 69, no. 6 (2014): 600–611.

59. Cowan, *More Work for Mother*, 38.

60. Donald W. Winnicott, "Environment," in *The Collected Works of D. W. Winnicott, vol. 11: Human Nature and The Piggle*, ed. Lesley Caldwell and Helen Taylor Robinson (New York: Oxford Academic, 2016).

61. Shaul Bar-Haim, *The Maternalists: Psychoanalysis, Motherhood, and the Welfare State* (Philadelphia: University of Pennsylvania Press, 2022).

62. Delap, *Knowing Their Place*.

63. Winnicott, "Environment."

64. Winnicott, "Environment."

65. Anne McClintock, *Imperial Leather: Race, Gender, and Sexuality in the Colonial Contest* (New York: Routledge, 1995), 86.

66. See Melanie Klein's *Love, Guilt, and Reparation* (New York: The Free Press, 1975). See also David Eng, "Colonial Object Relations," *Social Text* 34 (2016): 1–19.

67. Emily Green, "Melanie Klein and the Black Mammy: An Exploration of the Influence of the Mammy Stereotype on Klein's Maternal and Its Contribution to the 'Whiteness' of Psychoanalysis," *Studies in Gender and Sexuality* 19 (2018): 164–182; Amber Jamilla Musser, "Mammy's Milk and Absent Black Children," *Studies in Gender and Sexuality* 19 (2018): 188–190.

68. D. W. Winnicott, "Ego Distortion in Terms of True and False Self," *Maturational Processes and the Facilitating Environment: Studies in the Theory of Emotional Development* 64 (1960): 151.

69. "Piaget Kidnapping Memory," *Oxford Encyclopedia of the History of Modern Psychology*, ed. Wade E. Pickren (New York: Oxford University Press, 2022), 92.

70. G. Straker, "The Racialization of the Mind in Intimate Spaces: The 'Nanny' and the Failure of Recognition," *Studies in Gender and Sexuality* 13 (2012): 240–252.

71. Cornell University Home Economics Archive: Research, Tradition and History. Article unattributed, from *Good Housekeeping,* 1930.
72. Hall, "The Nursemaid," 20.
73. Hall, "The Nursemaid," 18.
74. Hall, "The Nursemaid," 23; Hall references two studies, one in Philadelphia and one in Chicago, both of which were forthcoming publications in 1931.
75. Article unattributed, from *Good Housekeeping*, vol. 114, no. 1 (1942): 127. Cornell University Home Economics Archive: Research, Tradition and History.
76. Article unattributed, from *Journal of Home Economics* 34, no. 3 (1942): 162. American Home Economics Association, Washington, Archive: Research, Tradition and History.
77. Martha Bensley Bruére, *Increasing Home Efficiency* (New York: Macmillan, 1923; reprint by Ulan Press, 2012), 201.
78. Isamu Noguchi, letter to E. F. McDonald Jr., 1946, MS_PROJ_057_004, Manuscript Collection, Project Files, Noguchi Archives.
79. Noguchi, letter to E. F. McDonald Jr., 1946.
80. "Art: Whitney Annual," *Time,* February 6, 1939. Review of Victoria and Albert Museum, Noguchi Radio Nurse: Museum no. W.16–2007.
81. For more on the specific panics attached to babysitters, see Miriam Forman-Brunell, *Babysitter: An American History* (New York: New York University Press, 2011).
82. Catherine A. Stewart, *Long Past Slavery: Representing Race in the Federal Writers' Project* (Chapel Hill: University of North Carolina Press, 2016), 13.
83. See Lilly Irani, "Difference and Dependence among Digital Workers: The Case of Amazon Mechanical Turk," *South Atlantic Quarterly* 114, no. 1 (2015): 225–234. See also Bernard Geoghegan, "Orientalism and Informatics: Alterity from the Chess-Playing Turk to Amazon's Mechanical Turk," *Exposition* 45 (2020); and Ranjodh Singh Dhaliwal, "The Cyber-Homunculus: On Race and Labor in Plans for Computation," *Configurations* 30, no. 4 (2022): 377–409.
84. Emilie Stoltzfus, *Citizen, Mother, Worker: Debating Public Responsibility for Childcare* (Chapel Hill: University of North Carolina Press, 2004), 20.
85. Miriam Forman-Brunell, *Babysitter: An American History* (New York: New York University Press, 2009).
86. "Never Leave Your Baby Alone," *Good Housekeeping*, July 1950, 100, 189. Cornell Home Economics Archive: Research, Tradition and History.
87. Donald W. Winnicott, "Primitive Emotional Development," in *The Collected Works of D. W. Winnicott, vol. 2: 1939–1945,* ed. Lesley Caldwell and Helen Taylor Robinson (New York: Oxford Academic, 2016).

88. Benjamin Spock, *The Common Sense Book of Baby and Child Care* (New York: Pocket Books, 1946). In the current version (the 10th edition), nanny has been removed altogether; see Spock, *Dr. Spock's Baby and Child Care*, 10th ed., revised and updated by Robert Needlman, MD (New York: Gallery Books, 2018).

89. Spigel, *Make Room for TV*.

90. Cowan, *More Work for Mother*, 142.

91. Sianne Ngai, *Theory of the Gimmick: Aesthetic Judgment and Capitalist Form* (Cambridge, MA: Belknap Press, 2020).

92. Gilbert Caluya, "Pride and Paranoia: Race and the Emergence of Home Security in Cold War America," *Continuum* 28, no. 6 (2014): 808–819.

93. Wendy Hui Kyong Chun, *Discriminating Data: Correlation, Neighborhoods, and the New Politics of Recognition* (Cambridge, MA: MIT Press, 2021).

94. But care is unevenly provided, and medical redlining is as rampant in postnatal healthcare as it is in maternal and neonatal care. While some women have a safe, healthy space to talk about the difficulties of early motherhood—while, yes, helpfully being monitored for signs of a postpartum psychiatric problem, in which medical oversight can quickly intersect with social services or the police—many women of color report declining to talk about postpartum problems because of their well-founded fear that police or social workers are more likely to take away their children.

95. Alice W. Newton and Andrea M. Vandeven, "Update on Child Maltreatment with a Special Focus on Shaken Baby Syndrome," *Current Opinion in Pediatrics* 17, no. 2 (April 2005): 246–251.

96. Adam Bryant, "In the Case of Working Mothers, Plenty of Judges," *New York Times*, November 16, 1997.

97. Edward Rothstein, "The Internet in the Courtroom Sparks More Clamor than Consensus," *New York Times*, November 24, 1997.

98. "Special Report: Judge Zobel's Order, a Summary," *BBC News*, November 10, 1997.

99. "Special Report: Judge Zobel's Order."

100. Penny Arevalo, "The Walls Have Eyes," *Los Angeles Times*, May 21, 1997.

101. Ian Ball, "Florida Baby-Shaking Case Sparks Rush to Buy Baby Monitors," *Sunday Telegraph*, October 19, 2003.

102. A. N. Guthkelch, "Problems of Infant Retino-Dural Hemorrhage with Minimal External Injury," *Houston Journal of Health Law & Policy* 12 (2012): 201–208.

103. Simone Browne, *Dark Matters: On Surveillance of Blackness* (Durham, NC: Duke University Press, 2015).

104. Susan Scheftel, "Why Aren't We Curious about Nannies?," *Psychoanalytic Study of the Child* 66 (2012): 251–278.

105. Daphne de Marneffe, *Maternal Desire: On Children, Love, and the Inner Life* (New York: Little Brown, 2004), 153–154.

106. "Diversity," Data USA: Childcare Workers, last accessed February 24, 2020, https://datausa.io/profile/soc/childcare-workers#demographics.

107. Knott, "Theorizing and Historicizing Mothering's Many Labours."

CHAPTER 2

1. See Katherine A. Ross, "Perpetuating 'Scientific Motherhood': Infant Feeding Discourse in *Parents* Magazine, 1930–2007," *Women & Health* 50, no. 3 (May 5, 2010): 297–311; Rima D. Apple, *Perfect Motherhood: Science and Childrearing in America* (New Brunswick, NJ: Rutgers University Press, 2006) and *Mothers and Medicine: A Social History of Infant Feeding, 1890–1950* (New Brunswick, NJ: Rutgers University, 1987); Jacquelyn S. Litt, "Taking Science to the Household: Scientific Motherhood in Women's Lives," in *Feminist Science Studies: A New Generation*, ed. Maralee Mayberry, Banu Subramaniam, and Lisa H. Weasel (New York: Routledge, 2020 [ebook]).

2. Eyal Ben-Ari, "'It's Bedtime' in the World's Urban Middle Classes: Children, Families and Sleep," in *Worlds of Sleep*, ed. Lodewijk Brunt and Brigitte Steger (Leipzig: Frank and Timme, 2008), 175.

3. See Christina Hardyment, *Dream Babies: Childcare Advice from John Locke to Gina Ford* (London: Quantro Press, 2007), 88–89.

4. Hardyment, *Dream Babies*, 89.

5. See Viviana Zelizer, *Pricing the Priceless Child* (Princeton, NJ: Princeton University Press, 1985).

6. Quoted in Hardyment, *Dream Babies*, 93.

7. For more on the history of formula and its relationship to breastfeeding, see Jessica L. Martucci, *Back to the Breast: Natural Motherhood and Breastfeeding in America* (Chicago: University of Chicago Press, 2015).

8. Peter N. Stearns, Perrin Rowland, and Lori Giarnella, "Children's Sleep: Sketching Historical Change," *Journal of Social History* 30, no. 2 (Winter 1996): 346.

9. Stearns, Rowland, and Giarnella, "Children's Sleep," 345–366.

10. Stearns, Rowland, and Giarnella, "Children's Sleep," 345–366.

11. Robin Bernstein, *Racial Innocence* (New York: New York University Press), 39.

12. See Karin Calvert, *Children in the House: The Material Culture of Early Childhood, 1600–1900* (Boston: Northeastern University Press, 1994).

13. William Wordsworth, "Michael: A Pastoral Poem," in *Lyrical Ballads, with Other Poems*, vol. 1 (Project Gutenberg, n.d.; originally published in 1800).

14. L. Emmett Holt, *The Care and Feeding of Children: A Catechism for the Use of Mothers and Children's Nurses* (New York: D. Appleton, 1910), https://www.gutenberg.org/files/15484/15484-h/15484-h.htm.

15. Holt, *The Care and Feeding of Children*.
16. Holt, *The Care and Feeding of Children*.
17. Stearns, Rowland, and Giarnella, "Children's Sleep," 358.
18. Elisabeth M. Yang, "Kingdoms of Babes: Home Nurseries in Turn-of-the-Century America," presented at the Cummings Center for the History of Psychology, October 27, 2022.
19. Stearns, Rowland, and Giarnella, "Children's Sleep," 345.
20. "Needs No Hand to Rock the Cradle: Electrical Device Perfected for Keeping the Cradle in Motion and for Amusing the Baby by the Play of Colored Lights," *Chicago Daily Tribune*, February 13, 1898, 41, https://search-proquest-com.libproxy.berkeley.edu/docview/172833010?accountid=14496. Thirty years later, in 1928, when roughly 50 percent of American residences were wired, the first electric crib patent was taken out by Herman H. Millard for a motorized cradle that simply featured automated rocking. Across the rest of the twentieth century, small changes to the idea of an automatic, rocking cradle resulted in subsequent patents. This invention had almost no market at the time of its creation, not only because electricity would not be in wide use for decades, but because the cradle—the non-automated version—had gone completely out of vogue during a time of significant revision to conceptions of child-rearing and care in middle-class homes. For the patents and descriptions of these cribs, see US Patent US1662754A, 1928; US Patent US2478445A, 1947; US Patent US2644958A, 1951; US Patent US3769641A, 1972; US Patent US4881285, 1988; US Patent US4881285, 1988.
21. US Patent US1662754A, Small beds for newborns or infants, e.g. bassinets or cradles with rocking mechanisms, 1928.
22. US Patent US2478445A, Small beds for newborns or infants, e.g. bassinets or cradles with rocking mechanisms, 1947.
23. US Patent US2644958A, Small beds for newborns or infants, e.g. bassinets or cradles with rocking mechanisms, 1951.
24. US Patent US3769641A, Small beds for newborns or infants, e.g. bassinets or cradles with rocking mechanisms, 1972.
25. US Patent US4881285, Small beds for newborns or infants, e.g. bassinets or cradles with rocking mechanisms, 1988.
26. "Hints about Health: Sleepless Nights," *Godey's Lady's Book*, December 1865, 448; "Health Department: Early Rising," *Godey's Lady's Book,* February 1859, 83; Annie Ramsey, "Nursery Furnishing," *Ladies' Home Journal,* February 1896, 24.
27. Dr. Fritz Talbot of Harvard Medical School gendered this: books for girls, radio for boys, as quoted in Stearns, Rowland, and Giarnella, "Children's Sleep," 351.
28. John B. Watson, *Psychological Care of Infant and Child* (New York: W. W. Norton, 1928).
29. Stearns, Rowland, and Giarnella, "Children's Sleep."

30. Carola Oßmer, "Normal Development: The Photographic Dome and the Children of the Yale Psycho-Clinic," *Isis* 111, no. 3 (2020): 515–541.
31. Deborah Blyth Dorshow, "An Alarming Solution: Bedwetting, Medicine, and Behavioral Conditioning in Mid-Twentieth Century America," *Isis* 101, no. 2 (June 2010): 314–315.
32. Dorshow, "An Alarming Solution ," 323.
33. B. F. Skinner "Baby in a Box," *Ladies' Home Journal*, October 1945.
34. Skinner, "Baby in a Box."
35. Skinner, "Baby in a Box."
36. Skinner, "Baby in a Box." As some examples, having to do laundry frequently was solved by making a temperature-controlled environment for their Deborah, which allowed her to move about in just a diaper, exercising vigorously, happily, and keeping herself entertained. The sheet was replaced with a roll of material that when soiled could be advanced, keeping the crib fresh and clean with just the turn of a crank. The crib also had a music box, which by seven months Deborah could operate herself—much like the first automated cradle.
37. Skinner, "Baby in a Box."
38. Deborah Skinner Buzan, "I Was Not a Lab Rat," letter to the *Guardian*, March 12, 2004. Skinner Buzan is indeed responding to the misstating of the facts of her childhood in the aftermath of the publication of Lauren Slater's *Opening Skinner's Box: Great Psychological Experiments of the Twentieth Century* (New York: W. W. Norton, 2004).
39. See for example, B. F. Skinner's prodigious correspondence in the 1960s with parents and gadgeteers. Papers of Burrhus Frederic Skinner Series, ca. 1932–1979, Air Crib, 1948 and 1968–1973, HUGFP 60.20, Box 1, Harvard University Archives, Countway Special Collections for the History of Medicine.
40. See Calvert, *Children in the House*.
41. Skinner, "Baby in a Box."
42. David Snyder, "Playroom," in *Cold War Hothouses: Inventing Postwar Culture, from Cockpit to Playboy*, ed. Beatriz Colomina, Ann Marie Brennan, and Jeannie Kim (Princeton, NJ: Princeton Architectural Press, 2004), 127.
43. Snyder, "Playroom."
44. "How to Stop Tantrums: Skinner Describes 'Operant Conditioning," *Harvard Law Review* 46, no. 5 (March 7, 1968). Papers of Burrhus Frederic Skinner Series, ca. 1932–1979, Air Crib, 1948 and 1968–1973, HUGFP 60.20, Box 1, Harvard University Archives, Countway Special Collections for the History of Medicine.
45. May Harrington Hall, "An Interview with 'Mr. Behaviorist': B. F. Skinner," *Psychology Today*, September 1967, 21–22. Papers of Burrhus Frederic Skinner Series, ca. 1932–1979, Air Crib, 1948 and 1968–1973, HUGFP 60.20, Box 1, Harvard University Archives, Countway Special Collections for the History of Medicine.

46. "B. F. Skinner: Father, Grandfather, Behavior Modifier," *Human Behavior* (January–February 1972): 15. Papers of Burrhus Frederic Skinner Series, ca. 1932–1979, Air Crib, 1948 and 1968–1973, HUGFP 60.20, Box 1, Harvard University Archives, Countway Special Collections for the History of Medicine.

47. See Papers of Burrhus Frederic Skinner Series, ca. 1932–1979, Air Crib, 1948 and 1968–1973, HUGFP 60.20, Box 1, Harvard University Archives, Countway Special Collections for the History of Medicine. As just one example, John L. MacIver Jr. to Skinner, April 10, 1948. Skinner was so used to this message that he had a standard response lamenting the delay of production and included home-manufacturing specifications.

48. B. F. Skinner to Mrs. Robert J. Walter, May 10, 1948. Papers of Burrhus Frederic Skinner Series, ca. 1932–1979, Air Crib, 1948 and 1968–1973, HUGFP 60.20, Box 1, Harvard University Archives, Countway Special Collections for the History of Medicine.

49. The Skinner's air crib met this exact fate in the popular press over and over again, long after it had become a relic of mid-century behaviorist history. The story of Deborah, the first air crib baby, went through a malicious game of telephone where authors and reporters decided that she had committed suicide after being a child experiment of her father's. Not only is she very much alive (and wrote an understandably angry letter to one such place that reported this misinformation), but she feels no harm was done her. While the cribs were never *mass*-produced, children were raised in the mis-remembered "Skinner box" and similarly labeled as "Skinner children." See the misreported story in Lauren Slater's *Opening Skinner's Box*. See Deborah Skinner Buzan's response to the book: "I Was Not a Lab Rat," letter to the *Guardian*, March 12, 2004.

50. Skinner to Dr. Phillup Blake, March 2, 1976, Papers of Burrhus Frederic Skinner Series, ca. 1932–1979, Air Crib, 1948 and 1968–1973, HUGFP 60.20, Box 1, Harvard University Archives, Countway Special Collections for the History of Medicine.

51. Rebecca Rego Barry, "Coney Island's Incubator Babies," *JSTOR Daily*, August 15, 2018, https://daily.jstor.org/coney-islands-incubator-babies. See also Janet Golden's *Babies Made Us Modern: How Infants Brought America into the Twentieth Century* (Cambridge: Cambridge University Press, 2018), especially chapter 1.

52. For instance, see Stephen F. Schell to B. F. Skinner on behalf of the Willowbrook state school neuro-endocrine research unity, April 23, 1968. Papers of Burrhus Frederic Skinner Series, ca. 1932–1979, Air Crib, 1948 and 1968–1973, HUGFP 60.20, Box 1, Harvard University Archives, Countway Special Collections for the History of Medicine.

53. Benjamin Spock, *The Common Sense Book of Baby and Child Care* (New York: Pocket Books, 1946), 1.

54. Ruth Gilbert, Georgia Salanti, Melissa Harden, and Sarah See, "Infant Sleeping Position and the Sudden Infant Death Syndrome: Systematic Review of Observational Studies and Historical Review of Recommendations from 1940 to 2002," *International Journal of Epidemiology* 34, no. 4 (August 2005): 874–887, https://doi.org/10.1093/ije/dyi088.

55. Salk to John Barbour, Salk Papers, M2399, Folder 14, Correspondence Files, Nicholas and Dorothy Cummings Center for the History of Psychology at the University of Akron.

56. Joost A. M. Meerloo, "Rhythm in Babies and Adults: Its Implications for Mental Contagion," *Archives of General Psychiatry* 5, no. 2 (1961): 169–175, https://jamanetwork.com/journals/jamapsychiatry/article-abstract/488027.

57. Various securitone advertisements, Salk Papers, M2399, Folder 14, Nicholas and Dorothy Cummings Center for the History of Psychology at the University of Akron.

58. M. Neal, "Vestibular Stimulation and Developmental Behavior in the Small Premature Infant," *Nursing Research Report* 3 (1967): 1–4.

59. Jean Liedloff, *The Continuum Concept: In Search for Happiness Lost* (New York: Da Capo Press, 1986).

60. For more on this history, see Martucci, *Back to the Breast*.

61. Cressida J. Heyes, "Reading Advice to Parents about Children's Sleep: The Political Psychology of a Self-Help Genre," *Critical Inquiry* 49, no. 2 (Winter 2023): 159. See also Sareeta Amrute, "Go the Fuck to Sleep: Well-Being, Welfare, and the Ends of Capitalism in US Discourses on Infant Sleep," *South Atlantic Quarterly* 115, no. 1 (January 2016): 125–148.

62. William Sears, *Creative Parenting: How to Use the New Continuum Concept to Raise Children Successfully from Birth to Adolescence* (New York: Dodd, Mead, 1983), 10, 34–35.

63. Sears, *Creative Parenting*, 171.

64. David Kaplan, *Silicon Boys and Their Valley of Dreams* (New York: Harper Collins, 2000), 140.

65. See Robert O. Self, *All in the Family: The Realignment of American Democracy since the 1960s* (New York: Hill and Wang, 2013).

66. This rhymes with other kinds of "media prophylaxis" that Dylan Mulvin argues are often restricted in use by those who might most benefit from them, that is, workers on the night shift who are withheld screen-shifting software. Dylan Mulvin, "Media Prophylaxis: Night Modes and the Prevention of Doing Harm," *Information and Culture* 53 (2018): 2.

67. Susanna Rosenbaum, *Domestic Economies: Women, Work, and the American Dream in Los Angeles* (Durham, NC: Duke University Press, 2017), 4.

68. As restfulness becomes collapsed with proper productive maternity, this recalled an earlier period in the early twentieth century of conceptions of wakefulness as masculine and industrious.

69. See Natasha Schüll's work on tracking, especially in her forthcoming book *Keeping Track*.

70. See Ruth Schwartz Cowan, *More Work for Mother: The Ironies of Household Technology from the Open Hearth to the Microwave* (New York: Basic Books, 1985).

71. Haytham El Wardany, *The Book of Sleep* (New York: Seagull Books, 2020).

CHAPTER 3

1. "Medicine: Frosted Children," *Time*, April 26, 1948.
2. "Medicine: Frosted Children."
3. For a vital and rigorous compendium of twentieth-century maternal blame and bad mother types, see Molly Ladd-Taylor and Lauri Umansky, eds., *"Bad" Mothers: The Politics of Blame in Twentieth-Century America* (New York: New York University Press, 1998).
4. For the intersection of animal research and attachment research, see Marga Vicedo, *The Nature and Nurture of Love: From Imprinting to Attachment in Cold War America* (Chicago: University of Chicago Press, 2013), especially chapters 2, 6, and 8.
5. For more on the influence of Watson's thinking in bad mother theories, see Ladd-Taylor and Umansky, *"Bad" Mothers*; see also Ann Hulbert, *Raising America: Experts, Parents, and a Century of Advice about Children* (New York: Vintage Books, 2004).
6. Anna Freud and Dorothy Burlingham, *Infants without Families: The Case for and against Residential Nurseries* (London: G. Allen and Unwin, 1943), and John Bowlby, "Maternal Care and Mental Health," *Bulletin of the World Health Organization* 3 (1951), as two classic examples of this work.
7. McLuhan problematically extends this binary of hot and cool to "backward countries" (cool) versus the "we" of the global north, which he terms "hot." Marshall McLuhan, *Understanding Media: The Extensions of Man* (Berkeley, CA: Gingko, 2013), 26.
8. Nicole Starosielski, "The Materiality of Heat Media," *International Journal of Communication* 8 (2014): 2504–2508, at 2505.
9. McLuhan, *Understanding Media*, 28.
10. As Nicole Starosielski points out, metaphorics of hot and cool are not exclusive to McLuhan; the discipline of film and media studies is rife with these metaphorics. See Starosielski. "The Materiality of Heat Media."
11. For more on McLuhan and race, see Armond Towns, "The (Black) Elephant in the Room: McLuhan and the Racial," *Canadian Journal of Communication* 44 (2019): 4, and Sarah Sharma's fantastic edited collection *Re-Understanding Media: Feminist Extensions of Marshall McLuhan* (Durham, NC: Duke University Press, 2022).
12. McLuhan, *Understanding Media*, 91
13. For more on Harry Harlow and these experiments, see Deborah Blum, *Love at Goon Park: Harry Harlow and the Science of Affection* (New York: Berkeley Books, 2012).
14. R. B. Cairns, "Attachment Behavior of Mammals," *Psychological Review* 73 (1966): 409–426; R. B. Cairns, Development, Maintenance, and Extinction of Social Attachment Behavior in Sheep," *Journal of Comparative Physiological Psychology* 62 (1966): 298–306.
15. Robert B. Cairns and Donald L. Johnson, "The Development of Interspecies Social Attachments," *Psychonomic Science* 2, no. 11 (1965): 337–338.

16. John Bowlby, *Attachment and Loss: Volume One* (London: Tavistock Institute of Human Relations, 1969), 180.
17. Later, in 1971, Harlow and Leo Kanner would go on to correspond on the question of inducing a *simulated* autistic and schizophrenic state in monkeys in order to deduce, perhaps once and for all, whether autism lay with the child or the mother. Harry Harlow, letter to Leo Kanner, June 10, 1971, American Psychiatric Association Foundation Archives, Leo Kanner Papers, Box 1000695, Folder Three, Autism and Childhood Schizophrenia, 1970–1973.
18. Leo Kanner, *In Defense of Mothers: How to Bring Up Children in Spite of the More Zealous Psychologists* (New York: Dodd, Mead, 1941), 45. For more on this contrast as it relates to the mothers of autistic children, see Jordyn Jack's *Autism and Gender: From Refrigerator Mothers to Computer Geeks* (Champaign: University of Illinois Press, 2004).
19. Leo Kanner, *In Defense of Mothers: How to Bring up Children in Spite of the More Zealous Psychologists* (New York: Dodd, Mead, 1941), 6.
20. Kanner, *In Defense of Mothers*.
21. Gregory Bateson, Don D. Jackson, Jay Haley, and John Weakland, "Toward a Theory of Schizophrenia," *Behavioral Science* 1, no. 4 (1956): 251–264.
22. For more on Bateson's work and its relationship to cybernetics, see Deborah Weinstein, *The Pathological Family: Postwar America and the Rise of Family Therapy* (Ithaca, NY: Cornell University Press, 2013), 47–48, and Bernard Dionysius Geoghegan, "The Family as Machine: Film, Infrastructure, and Cybernetic Kinship in Suburban America," *Grey Room* 66 (Winter 2017): 70–101.
23. Ellen Herman, *Kinship by Design: A History of Adoption in the Modern United States* (Chicago: University of Chicago Press, 2008), 255.
24. David M. Levy, *Maternal Overprotection* (New York: Columbia University Press, 1943), 161.
25. "Dr. David M. Levy, 84, a Psychiatrist, Dies," *New York Times*, March 4, 1977, https://www.nytimes.com/1977/03/04/archives/dr-david-m-levy-84-a-psychiatrist-dies-an-innovator-in-child.html.
26. "'Moms' Denounced as Peril to Nation," *New York Times*, April 28, 1945.
27. For more on the unprecedented surveillance project of the draft board, see David Serlin, "Crippling Masculinity: Queerness and Disability in US Military Culture, 1800–1945," *Gay and Lesbian Quarterly* 9, nos. 1–2 (2003): 149–179.
28. Edward A. Strecker, *Their Mother's Sons: The Psychiatrist Examines an American Problem* (Philadelphia: Lippincott, 1946), 6.
29. Strecker, *Their Mother's Sons*, 13. For more on "momism," see Philip Wylie, *A Generation of Vipers* (New York: Reinhart, 1942); See also W. Blatz, *Hostages to Peace: Parents and the Children of Democracy* (New York: Morrow, 1940). For more on this history, see Paula J. Caplan, "Mother-Blaming," in *"Bad" Mothers*, ed. Ladd-Taylor and Umansky; Barbara Ehrenreich and Deirdre English, *For Her Own Good: Two Centuries of the Experts' Advice to Women* (New

York: Penguin, 1978), 258–268; Rebecca Jo Plant, *Mom: The Transformation of Motherhood in Modern America* (Chicago: University of Chicago Press, 2010), 19–54; Mari Jo Buhle, *Feminism and Its Discontents* (Cambridge, MA: Harvard University Press, 2000), 125–164; and Anne Harrington, "Mother Love and Mental Illness: An Emotional History," *Osiris* 31 (2016): 94–115.

30. Weinstein, *The Pathological Family,* 25. For maternal dominance and its relationship to racist children, see Ruth Feldstein, *Motherhood in Black and White: Race and Sex in American Liberalism* (Ithaca, NY: Cornell University Press, 2000), 40–85. On Cold War containment strategies on the domestic front, see Elaine Tyler May, *Homeward Bound: American Families in the Cold War Era* (New York: Basic Books, 1988); on psychoanalysis, attachment theory, and World War II Britain, see Michal Shapira, *The War Inside: Psychoanalysis, Total War, and the Making of the Democratic Self in Postwar Britain* (Cambridge: Cambridge University Press, 2013).

31. Strecker, *Their Mother's Sons,* 30. See also the parents of Kanner's patients, who wrote of recovering the "missing links" as key to their child's "recovery" and "overcoming": Fred Grigspy, letter to Leo Kanner, September 6, 1970, American Psychiatric Association Foundation Archives, Leo Kanner Papers, Box 1000695, Folder Two, Autism, 1968–1977.

32. For more on this sequel and how Strecker pathologizes gay girls, see Jennifer Terry, "'Momism' and the Making of Treasonous Homosexuals," in *"Bad" Mothers,* ed. Ladd-Taylor and Umansky, 180–185.

33. Terry, "'Momism' and the Making of Treasonous Homosexuals," 180.

34. Strecker, *Their Mother's Sons,* 128.

35. See also Jules Gil-Peterson's *Histories of the Transgender Child* (Minneapolis: University of Minnesota Press, 2018), especially chapters 3 and 4.

36. Ellen Herman, *The Romance of American Psychology: Political Culture in the Age of Experts* (Berkeley: University of California Press, 1995), 278–279.

37. May, *Homeward Bound,* 110.

38. Lynn Spigel, *Make Room for TV: Television and the Family Ideal in Postwar America* (Chicago: University of Chicago Press, 1992), 42, 110. See also May's *Homeward Bound.*

39. For the long history of white women's subjugation of Black women who care for their children, see Stephanie E. Jones-Rogers, *They Were Her Property* (New Haven, CT: Yale University Press, 2019), 110–120. See also Jennifer L. Morgan, *Laboring Women: Reproduction and Gender in New World Slavery* (Philadelphia: University of Pennsylvania, 2004), especially chapter 4.

40. Feldstein, *Motherhood in Black and White,* 45–46; Daryl Michael Scott, *Contempt and Pity: Social Policy and the Image of the Damaged Black Psyche 1880–1996* (Chapel Hill: University of North Carolina Press, 1997), 72. Instead, Scott argues that Chicago School sociologists, most notably E. Franklin Frazier, thought of the Black family as disorganized due to social

upheaval, and on its way to reorganization. Scott carefully shows how Frazier's earlier work has been incorrectly connected to the Moynihan Report via the collapsing of political meanings as they differ in the context of Franklin's 1932 *The Negro Family in Chicago* and the 1965 Moynihan Report. Scott points to Frazier's understandings of the problems of matriarchy and the resultant issues with children as a problem of urban ecology, not race; white immigrants faced similar problems in the same urban spaces. Frazier located hope for the return of patriarchy in union membership and in World War II's economic impacts. In the postwar era, Frazier began to locate personality disorder in lower-class Black homes—which he partially located in the matriarchal structures found therein. Scott, *Contempt and Pity*, 41–55, 75–76.

41. Scott, *Contempt and Pity*, 74.

42. Scott, *Contempt and Pity*, 78–79, 106–108. See also Anne Harrington, *Mind Fixers: Psychiatry's Troubled Search for the Biology of Mental Illness* (New York: W. W. Norton, 2019). Abram Kardiner and Lionel Ovesey were psychoanalysts who particularly argued that Black single mothers were pathologically hot, dominating, and the source of emasculated Black masculinity. See also Abram Kardiner and Lionel Ovesey, *The Mark of Oppression: A Psychosocial Study of the American Negro* (New York: W. W. Norton, 1951).

43. The Moynihan Report, *The Negro Family: The Case for National Action*, 1965, Black Past, last accessed November 1, 2020, https://www.blackpast.org/african-american-history/moynihan-report-1965/.

44. The Moynihan Report, *The Negro Family*.

45. Kathleen Bond Stockton, *The Queer Child, or Growing Up Sideways* (Durham, NC: Duke University Press, 2009), 191.

46. Hortense J. Spillers, "Mama's Baby, Papa's Maybe: An American Grammar Book," *Diacritics* 17, no. 2 (1987): 66. For more on the Moynihan Report, see Tiffany Lethabo King, "Black 'Feminisms' and Pessimism: Abolishing Moynihan's Negro Family," *Theory & Event* 21, no. 1 (2018): 68–87.

47. Moynihan Report, *The Negro Family*, 66.

48. Kevin J. Mumford, "Untangling Pathology: The Moynihan Report and Homosexual Damage, 1965–1975." *Journal of Policy History* 24, no. 1 (2012): 53–73, https://doi.org/10.1017/S0898030611000376.

49. Spillers, "Mama's Baby, Papa's Maybe," 66–68.

50. Kanner's work is unfolding in the same period as Asperger's research in 1944, but his work would go unrecognized in the United States until the 1970s when Asperger syndrome is first included in the Diagnostic and Statistical Manual of Mental Disorders. For more on the longer history of autistic diagnoses, see Amit Pinchevski and John Durham Peters, "Autism and New Media: Disability between Technology and Society," *New Media & Society* 18, no. 11 (2016), and Chloe Silverman, *Understanding Autism: Parents, Doctors, and the History of a Disorder* (Princeton, NJ: Princeton University Press, 2012).

51. For more on this question of the importance of intelligence, see Christopher Sterwald and Jeffrey Baker, "Frosted Intellectuals: How Dr. Leo Kanner Constructed the Autistic Family," *Perspectives in Biology and Medicine* 62, no. 4 (Autumn 2019): 690–709.
52. Pinchevski and Peters, "Autism and New Media," 2507–2523.
53. "Frosted Children," *Time*, April 26, 1948, 77.
54. Jordynn Jack, *Autism and Gender: From Refrigerator Mothers to Computer Geeks* (Champaign: University of Illinois Press, 2004), 49.
55. Jonathan Metzl, *The Protest Psychosis: How Schizophrenia Became a Black Disease* (Boston: Beacon, 2009), xiii.
56. Metzl, *The Protest Psychosis*.
57. Metzl, *The Protest Psychosis*.
58. Lauretta Bender, "The Autistic Child," September 23, 1960, Box 10, Folder 3, Courtesy of the Brooklyn College Archives and Special Collections, the Papers of Dr. Lauretta Bender.
59. Groomer as quoted in David E. Simpson, dir., *Refrigerator Mothers* (Kartemquin Films, 2002).
60. Simpson, *Refrigerator Mothers*.
61. Ruth Schwartz Cowan, *More Work for Mother: The Ironies of Household Technology from the Open Hearth to the Microwave* (New York: Basic Books, 1985), 203.
62. Cowan, *More Work for Mother*, 195.
63. Cowan, *More Work for Mother*.
64. Cowan, *More Work for Mother*, 196.
65. Ironically, advocacy by these parent organizations who organized in part around rejecting refrigerator mother rhetoric had a large hand in moving toward applied behavior analysis as the dominant autism treatment paradigm in the United States. Laura Schreibman, *The Science and Fiction of Autism* (Cambridge, MA: Harvard University Press, 2007), 84.
66. Schreibman, *The Science and Fiction of Autism*.
67. For the longer history of "emotionally disturbed children" and their treatment in institutions, see Deborah Blythe Doroshow, *Emotionally Disturbed: A History of Caring for America's Troubled Children* (Chicago: University of Chicago Press, 2019).
68. Silverman, *Understanding Autism*, 83
69. For a longer discussion of Bettelheim's influences, see Silverman, *Understanding Autism*, 66–70.
70. Ehrenreich and English, *For Her Own Good*, 205. For more on Betty Friedan and her readers, see Rebecca Jo Plant, *Mom: The Transformation of Motherhood in Modern America* (Chicago: University of Chicago Press, 2010), 146–176.

71. For more on "Joey," cybernetics, and schizophrenia, see Jeffrey Sconce, *The Technical Delusion: Electronics, Power, Insanity* (Durham, NC: Duke University Press, 2019), 224–235; Pinchevski and Peters, "Autism and New Media"; Sungook Hong, "Joey the Mechanical Boy, Revisited," *e-flux* (March 2018), https://www.e-flux.com/architecture/superhumanity/179228/joey-the-mechanical-boy-revisited.

72. For more on power and the case study, see Lauren Berlant, "On the Case," *Critical Inquiry* 33, no. 4 (2007): 663–672.

73. Bruno Bettelheim, "Joey: A 'Mechanical Boy,'" *Scientific American* 200, no. 3 (1959): 117.

74. Bettelheim, "Joey," 118.

75. Bettelheim, "Joey," 124.

76. Bettelheim, "Joey," 126.

77. Bettelheim, "Joey," 119.

78. Bettelheim, "Joey," 127.

79. Donna Hermawati, Farid Agung Rahmadi, Tanjung Ayu Sumekar, and Tri Indah Winarni, "Early Electronic Screen Exposure and Autistic-Like Symptoms," *Intractable & Rare Diseases Research* 7, no. 1 (2018): 69–71, doi: 10.5582/irdr.2018.01007

80. See Pinchevski and Peters, "Autism and New Media"; Alexander Galloway, *Laruelle: Against the Digital* (Minneapolis: University of Minnesota Press, 2014); James J. Hodge, "The Subject of Always-On Computing: Thomas Ogden's 'Autistic-Contiguous Position' and the Animated GIF," *Parallax* 26, no. 1 (2020): 65–75. See also Jeff Nagy, *Watching Feeling: Emotional Data from Cybernetics to Social Media*, Stanford University, Department of Communication, Dissertation, 2022, and Meryl Alper, *Kids across the Spectrums: Growing Up Autistic in the Digital Age* (Cambridge, MA: MIT Press, 2023).

81. Haim Ginott and Alice Ginott, *Between Parent and Child* (New York: Harmony, 1969).

82. Attachment parenting is not the same thing as a praxis of attachment theory, though they share some features (notably, attachment parenting is a much more extreme revision; if Bowlby and Ainworth think of the good mother as a safe base from which a child explores their world, coming and going, attachment parenting advocates a continuous joining of the dyad). Attachment parenting began to pervade parenting discourse in the 1970s and into the 1980s. An intensification of Dr. Spock's philosophy, attachment parenting came into its own in 1975 through Jean Liedloff's term "continuum concept." See chapter 2. For more on this history, see Jessica L. Martucci, *Back to the Breast: Natural Motherhood and Breastfeeding in America* (Chicago: University of Chicago Press, 2015).

83. See R. Shoup, R. M. Gonyea, and G. D. Kuh, "Helicopter Parents: Examining the Impact of Highly Involved Parents on Student Engagement and Educational Outcomes," presented at the 49th Annual Forum of the Association for Institutional Research, June 2009, Atlanta, Georgia.

84. The most notorious example of this is Amy Chua, *Battle Hymn of the Tiger Mother* (New York: Penguin Books, 2011).

85. Malcolm Harris, *Kids These Days: Human Capital and the Making of Millennials* (New York: Little, Brown, 2017), 6. For more on diagnosis of children in contemporary times, see Linda M. Blum, *Raising Generation RX: Mothering Kids with Invisible Disability in an Age of Inequality* (New York: New York University Press, 2015); Ara Francis, *Family Trouble: Middle-Class Parents, Children's Problems, and the Disruption of Everyday Life* (New Brunswick, NJ: Rutgers University Press, 2015); as well as Rayna Rapp and Faye Ginsburg, "Enabling Disability: Rewriting Kinship, Reimagining Citizenship," *Public Culture* 13, no. 3 (2001): 533–556.

86. For one example of statistical studies that perform this critique, see Matthias Doepke and Fabrizio Zilibotti, *Love, Money, and Parenting: How Economics Explains the Way We Raise Our Kids* (Princeton, NJ: Princeton University Press, 2019); for a study on college-educated mothers staying home (in the UK context), see Shani Orgad, *Heading Home: Motherhood, Work, and the Failed Promise of Equality* (New York: Columbia University Press, 2019), especially part II, chapter 3.

87. Hara Estroff Marano, *A Nation of Wimps: The High Cost of Invasive Parenting* (New York: Crown Archetype, 2008).

88. For more on surveillance and datafication of children, see Veronica Barassi, *Child Data Citizen* (Cambridge, MA: MIT Press, 2020); see also Hannah Zeavin, "Family Scanning," *Real Life Magazine*, February 22, 2021.

89. For an example of this literature, see L. M. Padilla-Walker and L. J. Nelson, "Black Hawk Down? Establishing Helicopter Parenting as a Distinct Construct from Other Forms of Parental Control during Emerging Adulthood," *Journal of Adolescence* 35 (2012): 1177–1190; R. L. Munich and M. A. Munich, "Overparenting and the Narcissistic Pursuit of Attachment," *Psychiatric Annals* 39 (2009): 227–235.

CHAPTER 4

1. 2021 Assembly Bill 627, Wisconsin State Assembly.

2. 2021 Assembly Bill 627, Wisconsin State Assembly.

3. Wisconsin Incarceration Trends, Vera Institute of Justice, December 2019. For a definitive study of prison expansion in this period, see Ruth Wilson Gilmore, *Golden Gulag: Prisons, Surplus, Crisis, and Opposition in Globalizing California* (Berkeley: University of California Press).

4. For more on algorithmic racism and incarceration, see the work of Virginia Eubanks, Wendy Chun, and Ruha Benjamin.

5. The use of algorithmic sentencing was famously taken to the supreme court of Wisconsin after Loomis appealed his sentence. The court upheld the use of COMPAS; the Supreme

Court of the United States declined to hear the case, thus setting the Wisconsin ruling as de facto national statute. See *Loomis v. Wisconsin*, 881 N.W.2d 749 (Wis. 2016).

6. Geof Bowker, "How to Be Universal: Some Cybernetic Strategies, 1943–70," *Social Studies of Science* 23, no. 1 (1993): 107–127, http://www.jstor.org/stable/285691.

7. Jackie Wang, *Carceral Capitalism* (Cambridge, MA: Semiotext(e), 2018).

8. See Dorothy Roberts, *Shattered Bonds: The Color of Child Welfare* (New York: Basic Books, 2000).

9. Mary Pat Brady, *Scales of Captivity: Racial Capitalism and the Latinx Child* (Durham, NC: Duke University Press, 2022), 11.

10. Every Second Project, "The Impact of the Incarceration Crisis on America's Families," Cornell University, 2021.

11. Dorothy Roberts, *Shattered Bonds: The Color of Child Welfare* (New York: Basic Books, 2000); Mical Raz, *Abusive Policies: How the American Child Welfare System Lost Its Way* (Chapel Hill: University of North Carolina Press, 2020).

12. Susan Hatters Friedman, Aimee Kaempf, and Sarah Kauffman, "The Realities of Pregnancy and Mothering while Incarcerated," *Journal of the American Academy of Psychiatry and the Law* 48, no. 3 (2020) 365–375, https://doi.org/10.29158/JAAPL.003924-20.

13. For more on birthing and reproduction in prison, see Carolyn Suffrin, *Jail Care: Finding the Safety Net for Women behind Bars* (Berkeley: University of California Press, 2017). For writings of women who have given birth in prison, including at Bedford Hills, see Bell Chevigny, ed., *Doing Time: Twenty-Five Years of Prison Writing* (New York: Arcade Publishing, 1999), and Rickie Solinger, Paula C. Johnson, Martha L. Raimon, Tina Reynolds, and Ruby C. Tapia, eds., *Interrupted Life: Experiences of Incarcerated Women in the United States* (Berkeley: University of California Press, 2000).

14. For more on life in women's prisons in the United States, including parent-infant programs and mothering, see Solinger, Johnson, Raimon, Reynolds, and Tapia, eds., *Interrupted Life*. For more on life after prison, including parenting, see Bruce Western, *Homeward: Life in the Year after Prison* (New York: Russell Sage Foundation, 2018). For more on fathers specifically, see Tassell McKay, Megan Comfort, Christine Lindquist, and Anupa Bir, *Holding On: Family and Fatherhood during Incarceration and Reentry* (Berkeley: University of California Press, 2019). See also Joyce A. Arditti, *Parental Incarceration and the Family: Psychological and Social Effects of Imprisonment on Children, Parents, and Caregivers* (New York: New York University Press, 2016), and Megan Comfort, *Doing Time Together: Love and Family in the Shadow of the Prison* (Chicago: University of Chicago Press, 2008).

15. Caitlin Rosenthal, *Accounting for Slavery: Masters and Management* (Cambridge, MA: Harvard University Press, 2018); Walter Johnson, *Soul by Soul: Life inside the Antebellum Slave Market* (Cambridge, MA: Harvard University Press, 1999); Jennifer L. Morgan, *Reckoning with Slavery: Gender, Kinship, and Capitalism in the Early Black Atlantic* (Durham, NC: Duke University Press, 2021). See also Sarah Haley, *No Mercy Here: Gender, Publishment, and the*

Making of Jim Crow Modernity (Chapel Hill: University of North Carolina Press, 2016), and Wendy Gonaver, *The Peculiar Institution and the Making of Modern Psychiatry 1840–1880* (Chapel Hill: University of North Carolina Press, 2019).

16. Joy James, "The Womb of Western Theory: Trauma, Timetheft, and the Captive Maternal," in *Challenging the Punitive Society: The Carceral Notebooks*, vol. 12, ed. Bernard Harcourt et al. (New York: Publishing Data Management, 2016), 255.

17. Wendy Kline, *Building a Better Race: Gender, Sexuality, and Eugenics from the Turn of the Century to the Baby Boom* (Berkeley: University of California Press, 2001).

18. See Roberts, *Shattered Bonds*.

19. James, "The Womb of Western Theory," 256.

20. Estelle B. Freedman, *Their Sisters' Keepers: Women's Prison Reform in America, 1830–1930* (Ann Arbor: University of Michigan Press, 1984).

21. Ilse Denisse Catalan, "Making Mothers," in *Incarcerated Women: A History of Struggles, Oppression, and Resistance in American Prisons*, ed. Erica Rhodes Hayden (Lanham, MD: Lexington Books, 2017).

22. Malcolm X, *The Autobiography of Malcolm X* (New York: Ballantine Books, 1965).

23. In addition to revering the line of inquiry into the relationship between computation and the carceral, the case of the mother-baby prison unit, its logic and my subsequent argument fall into a sub-tradition of the history of sciences and the history of technology. In this tradition, historians such as Fred Turner, Mara Mills, Bernard Geoghegan, and Deborah Weinstein and media theorists Virginia Eubanks, Wendy Chun, and Jackie Wang have restaged the origins of cybernetics. As Wang has it, cybernetics is "an ideology of management, self-organization, rationalization, control, automation, and technical certitude, giving rise after 1945" to point to new understandings of the formation of digital media. Jackie Wang, "This is a Story about Nerds and Cops: PredPol and Algorithmic Policing," *e-flux* 87 (December 2017).

24. Bernard Geoghegan, *Code* (Durham, NC: Duke University Press, 2022).

25. Nikolas Rose, *Governing the Soul: The Shaping of the Private Self* (New York: Free Association Books, 1998).

26. For the history of interwar psychology as it presages the maternal deprivation hypothesis, see Bican Polat, "Mental Hygiene, Psychoanalysis, and Interwar Psychology: The Making of the Maternal Deprivation Hypothesis," *Isis* 112, no. 2 (2021): 266–290. See also Lisa Cartwright, "'Emergencies of Survival': Moral Spectatorship and the 'New Vision of the Child' in Postwar Child Psychoanalysis." *Journal of Visual Culture* 3, no. 1 (2004): 35–38.

27. For more on attachment theory and its status in the Cold War era, see Marga Vicedo's *The Nature & Nurture of Love: From Imprinting to Attachment in Cold War America* (Chicago: University of Chicago Press, 2016). See also Elaine Tyler May, *Homeward Bound: American Families in the Cold War Era* (New York: Basic Books, 1988).

28. John Bowlby, "Forty-Four Juvenile Thieves: Their Characters and Home Lives," *International Journal of Psychoanalysis* 25 (1944): 19–53; D. W. Winnicott, too, in this period and after focused on the intertwined nature of separation or deprivation and delinquency in children, from 1939 to 1970. See *Deprivation and Delinquency* (New York: Basic Books, 1978).

29. Bowlby, "Forty-Four Juvenile Thieves."

30. D. W. Winnicott to John Bowlby, *The Spontaneous Gesture: The Selected Letters of D. W. Winnicott* (New York: Routledge, 1987), 65–66.

31. Frank C. P. van der Horst, Lenny van Rosmalen, and René van der Veer, "John Bowlby's Critical Evaluation of the Work of René Spitz," *History of Psychology* 22, no. 2 (2019): 205–208.

32. René Spitz, "Hospitalism," *Psychoanalytic Study of the Child* 1, no. 1 (1945): 53–74.

33. In some Spitz scholarship, it is presumed that the mothers in the penal colony had access to their children *full time*. This was evidentially not the case when consulting the archives of Westfield State Farm; visitation was clearly highly regulated.

34. René Spitz Papers, M2137, Folder 83, the Nicholas and Dorothy Cummings Center for the History of Psychology at the University of Akron.

35. John Bowlby and James Robertson, "A Two-Year-Old Goes to Hospital," *Proceedings of the Royal Society of Medicine* 46, no. 6 (1953): 425–427. See the René Spitz Papers, M2111, Folder 13, and M2137, Folder 87, for examples of the questionnaires given to mothers.

36. This was by no means Spitz's only film; he applied for copyright for seven other films in this period. *Grief* remains his most well-known film. René Spitz Papers, M2137, Folder 87. As Lisa Cartwright documents, the film had a long life in both psychology classrooms and in child wellness policies: "Spitz went to great lengths to make his message a global one, trying for example to get the WHO to act as global distributor of his films when they agreed to acquire some prints. Invoices and request letters in the Spitz collection show that the film circulated broadly among rural child guidance centers to the Jewish Board of Guardians, to intellectuals (including Bruno Bettelheim and Margaret Mead), to the American Medical Association, the US Public Health Service and the National Institutes of Mental Health. *Grief* continued to be used in psychology classes at least into the late 1990s." Cartwright, "'Emergencies of Survival,'" 45.

37. René Spitz Papers, M2137, Folder 83.

38. René Spitz, *Grief: A Peril in Infancy* (video, 1947), René Spitz Papers. See also Rachel Weitzenkorn, "Boundaries of Reasoning in Cases: The Visual Psychoanalysis of René Spitz," *History of the Human Sciences* 33, nos. 3–4 (2020), https://doi.org/10.1177/0952695120908491.

39. René Spitz Papers, M2137, Folder 83.

40. Katie Joice, "Mothering in the Frame: Cinematic Microanalysis and the Pathogenic Mother, 1945–67," *History of the Human Sciences* 34, no. 5 (December 2021): 105–131; Anne Harrington, "Mother Love and Mental Illness: An Emotional History," *Osiris* 31, no. 1 (July 1, 2016): 94–115.

41. Felix E. Rietmann, "Seeing the Infant: Audiovisual Technologies and the Mind Sciences of the Child," PhD diss., Princeton University, 2018, 21.

42. Cartwright, "'Emergencies of Survival,'" 37.

43. Weitzenkorn, "Boundaries of Reasoning in Cases," 7.

44. Mary Ann Doane, *Bigger than Life* (Durham, NC: Duke University Press, 2022).

45. René Spitz Papers, M2137, Folder 83. Several scholars working in Spitz's archives have also identified it as such, and again, this has remained in the *footnotes*.

46. Tonia Sutherland, "The Carceral Archive: Documentary Records, Narrative Construction, and Predictive Risk Assessment," *Journal of Cultural Analytics* 4, no. 1 (June 4, 2019).

47. The videography of the clinic has given way to the training of algorithms on mass facial recognition datasets, neuroimaging, and other high-tech solutions to the problem of screening mothers and their children; these formalized diagnostic criteria, and their pathological norming, have extended their reach, becoming a dragnet that further encodes racial and classed biases elaborated nearly fifty years ago.

48. For more on the history of the women's reformatory in New York in this period, see Nicole Hahn Rafter, *Partial Justice: Women in State Prisons 1800–1935* (Boston: Northeastern University Press, 1985), 44–80. See also Cheryl D. Hicks, *Talk with You like a Woman: African American Women, Justice, and Reform in New York, 1890–1935* (Chapel Hill: University of North Carolina Press, 2010).

49. Cyndi Banks, *Women in Prison: A Reference Handbook* (Santa Barbara, CA: ABC-CLIO, 2003). Later, in 1964, the incarcerated mothers at Westfield State were monitored by a social worker in a post established in conjunction with the New York City Protestant Episcopal Mission Society, which conducted studies on the effects of postpartum counseling on recidivism rates. Ilse Denisse Catalan, "Making Mothers," in *Incarcerated Women: A History of Struggles, Oppression, and Resistance in American Prisons,* ed. Erica Rhodes Hayden (Lanham, MD: Lexington Books, 2017), 132.

50. Nancy Campbell, *Using Women: Gender, Drug Policy, and Social Justice* (New York: Routledge, 2000), 102.

51. See Freedman, *Their Sisters' Keepers*. For more on mid-century New York and the carceral apparatus, see Carl Suddler, *Presumed Criminal: Black Youth and the Justice System in Postwar New York* (New York: New York University Press, 2019). For more on the history of the intertwined nature of asylum and prison in this period, see Anne E. Parsons, *From Asylum to Prison: Deinstitutionalization and the Rise of Mass Incarceration after 1945* (Chapel Hill: University of North Carolina Press, 2018).

52. Catalan, "Making Mothers," 132.

53. In the 1930s and 1940s, conceptions of maternal fitness (i.e., who is and can be a good mother) were very much in flux between Progressive-era positive and negative eugenicist

understandings and the panics around improper mothering that attended the Cold War. Across this period, the work of separating an unfit mother from her child became more deeply and legally entrenched, away from the disruption of the family. And this centered on two intertwined institutions: the prison and the Child Welfare Bureau.

54. Westfield State Farm, Nursery Day Books, 1961–1963, B1575, New York State Archives.

55. Westfield State Farm, Nursery Day Books.

56. Catalan, "Making Mothers," 137.

57. Westfield State Farm, Nursery Day Books.

58. Ellen Herman, *Kinship by Design: A History of Adoption in the Modern United States* (Chicago: University of Chicago Press, 2008), 259.

59. Wendi Klein, *Building a Better Race: Gender, Sexuality, and Eugenics from the Turn of the Century to the Baby Boom* (Berkeley: University of California Press), 2005; Nicole Hahn Rafter, *Partial Justice: Women in State Prisons 1800–1935* (Boston: Northeastern University Press, 1985), 44–80. See also Hicks, *Talk with You like a Woman*.

60. See S. Curtis, "'Tangible as Tissue': Arnold Gesell, Infant Behavior, and Film Analysis," *Science in Context* 24, no. 3 (2011): 417–442. For more on mid-century maternal research, see Joice, "Mothering in the Frame," 105–131.

61. Felix Rietmann, "Of Still Faces and Micro-Plots: Audiovisual Narration in Infant Mental Health," in *Narrative Structure and Narrative Knowing in Medicine and Science,* ed. Martina King and Tom Kindt (Berlin: de Gruyter, 2023) and Lisa Cartwright, *Moral Spectatorship: Technologies of Voice and Affect in Postwar Representations of the Child* (Durham, NC: Duke University Press, 2008). See also Lauren B. Adamson and Janet E. Prick, "The Still Face: A History of a Shared Experimental Paradigm," *Infancy* 4, no. 4 (2003): 451–473; Karen Adolph, "Video as Data: From Transient Behavior to Tangible Recording," *Association for Psychological Science* 29, no. 3 (2016): 23–25.

62. Vicedo, *The Nature & Nurture of Love*.

63. Rietmann. "Seeing the Infant," 61.

64. Mary D. Salter Ainsworth, ed., *Patterns of Attachment: A Psychological Study of the Strange Situation* (Hillsdale, NJ: Lawrence Erlbaum Associates, 1978); Mary Ainsworth and Silvia M. Bell, "Attachment, Exploration, and Separation: Illustrated by the Behavior of One-Year-Olds in a Strange Situation," *Child Development* 41, no. 1 (1970): 49–67.

65. Seth Barry Watter, "Scrutinizing: Film and the Microanalysis of Behavior," *Grey Room* 66 (2017): 32–69. For other psychological research on the family, Bateson, and the use of film, see Bernard Dionysius Geoghegan, "The Family as Machine: Film, Infrastructure, and Cybernetic Kinship in Suburban America," *Grey Room* 66 (2017): 70–101. For more on microanalysis as evidentiary regime, see J. Canales, *A Tenth of a Second: A History* (Chicago: University of Chicago Press, 2009); H. Landecker, "Microcinematography and the History

of Science and Film," *Isis* 97, no. 1 (2006): 121–132. For more on microanalysis in psychotherapy, see M. Lempert, "Fine-Grained Analysis: Talk Therapy, Media, and the Microscopic Science of the Face-to-Face," *Isis* 110, no. 1 (2019): 24–47.

66. In *Legal Spectatorship: Slavery and the Visual Culture of Domestic Violence* (Durham, NC: Duke University Press, 2022), Kelli Moore argues that "slavery's afterlife permeates throughout audio-visual culture" in the contemporary courtroom (focusing on domestic violence cases). Moore's genealogy of this visual culture is crucial to understanding the modes of knowing and looking that attachment theory enters into, especially in cases where parental rights are being terminated at the behest of co-parents.

67. Bernard E. Harcourt, *Against Prediction: Profiling, Policing, and Punishing in an Actuarial Age* (Chicago: University of Chicago Press, 2006), 39. See also Colin Koopman, *How We Became Our Data: A Genealogy of the Informational Person* (Chicago: University of Chicago Press, 2019).

68. For more on the longer history of mandatory minimums and drug-related offenses, see the Data & Civil Rights website, 2014–2015, datacivilrights.org. See also Angèle Christin, Alex Rosenblat, and danah boyd, "Data and Civil Rights: Courts and Predictive Algorithms," *Data & Society* (2015). This table became an advisory force rather than mandatory in 2005.

69. The Sentencing Project, "Incarcerated Women and Girls," 2019. See also E. A. Carson, "Prisoners in 2019," Bureau of Justice Statistics, Washington, DC; A. Zeng, "Jail Inmates in 2018," Bureau of Justice Statistics, Washington, DC; D. Kaeble and M. Alper, "Probation and Parole in the United States, 2017–2018," Bureau of Justice Statistics, Washington, DC.

70. H.R.5773—98th Congress (1984). See Christin, Rosenblat, and boyd, "Data and Civil Rights."

71. For more on the widespread implication of the mandatory minimum, see Harcourt, *Against Prediction*.

72. Public Law 105-89—November 19, 1997, 111 STAT. 2115

73. See Roberts, *Shattered Bonds*.

74. Moore was interviewed by *People* magazine during the trial, and a film on her court case, "Love Child," was released in 1982.

75. Wainwright v. Moore, 374 So.2d 586 (1979)

76. West's F.S.A. § 944.24; 30 West's F.S.A. Rules of Civil Procedure, rule 1.210(b).

77. Sandra Hinson, "Terry Jean Moore Fights to Keep Her Baby in a Florida Prison," *People*, May 21, 1979.

78. Hinson, "Terry Jean Moore Fights to Keep Her Baby."

79. Hinson, "Terry Jean Moore Fights to Keep Her Baby."

80. Hinson, "Terry Jean Moore Fights to Keep Her Baby."

81. Roberts, *Shattered Bonds*; Priya Kandaswamy, *Domestic Contradictions: Race and Gendered Citizenship from Reconstruction to Welfare Reform* (Durham, NC: Duke University Press, 2021); Mical Raz, *Abusive Policies: How the American Child Welfare System Lost its Way* (Chapel Hill: University of North Carolina Press, 2020).

82. H.R.867—Adoption and Safe Families Act of 1997.

83. Re E.A.P., 944 A.2d 79 (2008).

84. Leeper v. Leeper, 21 So.3d 1006 (2009).

85. RE SCARLET W. et al. No. W2020–00999-COA-R3-PT.

86. 66 Misc.3d 1217(A), 120 N.Y.S.3d 726, 2020 N.Y. Slip Op. 50116(U).

87. 66 Misc.3d 1217(A), 120 N.Y.S.3d 726, 2020 N.Y. Slip Op. 50116(U).

88. For more on the uprisings, see Victoria Law, *Resistance behind Bars: The Struggles of Incarcerated Women* (New York: PM Press, 2009).

89. Even by the late 1980s, many studies had shown that kin and connection produce lower rates of recidivism. For a literature overview from this period, see C. F Hairston, "Family Ties During Imprisonment: Do They Influence Future Criminal Activity?," *Federal Probation* 52, no. 1 (March 1988): 48–52.

90. L. E. Glaze and L. M. Maruschak, "Parents in Prison and Their Minor Children," US Department of Justice, Bureau of Justice Statistics, Special Report. NCJ 222984, 2008; C. J. Mumola, "Incarcerated Parents and Their Children," US Department of Justice, Bureau of Justice Statistics, Special Report. NCJ 182335, 2000.

91. M. W. Byrne, L. S. Goshin, and S. S. Joesti, "Intergenerational Transmission of Attachment for Infants Raised in a Prison Nursery," *Attachment & Human Development* 12, no. 4 (2010): 375–393, doi: 10.1080/14616730903417011.

92. Joice, "Mothering in the Frame," 130, note 3.

93. Byrne, Goshin, and Joesti, "Intergenerational Transmission of Attachment for Infants Raised in a Prison Nursery"; see also Claire Powell, Lisa Marzano, and Karen Ciclitira, "Mother-Infant Separations in Prison: A Systematic Attachment-Focused Policy Review," *Journal of Forensic Psychiatry & Psychology* 28 (2016): 1–16. Powell, Marzano, and Ciclitira studied British documents, which demonstrate widespread implicit recourse, as well as direct reference, to attachment theory in UK prison nurseries and their policies.

94. For more on psychiatric power in the contemporary carceral context, see Anthony Ryan Hatch, *Silent Cells: The Secret Drugging of Captive America* (Minneapolis: University of Minnesota Press, 2019).

95. For more on the contemporary life inside the prison nursery at Bedford Hills, see Sarah Yager, "Prison Born," *Atlantic*, July 2015.

96. *Prisons*. Canada: CLOG, 2014.

CHAPTER 5

1. "How 'It's 10 pm...' Began," *Daily News*, October 22, 1985, 76.
2. Henry Goldschmidt, *Race and Religion among the Chosen People of Crown Heights* (New Brunswick, NJ: Rutgers University Press, 2006), 94.
3. For an extended consideration of the history of juvenile "delinquency" in the justice system in New York at this time, see Carl Suddler, *Presumed Criminal: Black Youth and the Justice System in Postwar New York* (New York: New York University Press, 2020).
4. "How 'It's 10 pm...' Began," 76.
5. "How 'It's 10 pm...' Began."
6. "How 'It's 10 pm...' Began."
7. For a history of early media research and reform as regards the radio, see Josh Sheppard, *Shadow of the New Deal: The Victory of Public Broadcasting* (Champaign: University of Illinois Press, 2023).
8. Garth S. Jowett, Ian C. Jarvie, and Kathryn H. Fuller, *Children and the Movies: Media Influence and the Payne Fund Controversy* (New York: Cambridge University Press, 1996).
9. Otniel E. Dror, "The Affect of Experiment: The Turn to Emotions in Anglo-American Physiology, 1900–1940," *Isis* 90, no. 2 (1999): 205–237.
10. Brenton J. Malin, "Mediating Emotion: Technology, Social Science, and Emotion in the Payne Fund Motion Picture Studies," *Technology and Culture* 50, no. 2 (April 2009): 366–390.
11. Deborah Coon, "Standardizing the Subject: Experimental Psychologists, Introspection, and the Quest for a Technoscientific Ideal," *Technology and Culture* 34, no. 4 (1993): 757–783.
12. William J. Buxton, "From Park to Cressey: Chicago Sociology's Engagement with Media and Mass Culture," in *The History of Media and Communication Research: Contested Memories*, ed. David Park and Jefferson Pooley (New York: Peter Lang, 2008, 348).
13. Buxton, "From Park to Cressey," 346.
14. The thirteenth study, originally part of the Payne Fund, was initially suppressed and only found some sixty years later by film historians Garth S. Jowett, Ian C. Jarvie, and Kathryn H. Fuller; see their *Children and the Movies*.
15. Annette Kuhn and Guy Westwell, "Payne Fund Studies," in *A Dictionary of Film Studies* (New York: Oxford University Press, 2012).
16. Kuhn and Westwell, "Payne Fund Studies."
17. Sterling North, "A National Disgrace," *Childhood Education* 17, no. 1 (January 1940): 56 [reprinted in *Chicago Daily News*, 1940].
18. Kevin Duong, "Broke Psychoanalysis," *Parapraxis* 3 (December 2023): 56.
19. Duong, "Broke Psychoanalysis."

20. Wertham, quoted in Duong, "Broke Psychoanalysis."
21. See Elizabeth Ann Danto, *Freud's Free Clinics: Psychoanalysis and Social Justice, 1918–1938* (New York: Columbia University Press, 2007).
22. Duong, "Broke Psychoanalysis."
23. New York Public Library, Archives and Manuscripts, Lafargue Clinic Records, Box 5, Folder 17. Script for "The House I Enter: The Story of an American Doctor."
24. For more on Wertham, see Anne Harrington, *Mind Fixers: Psychiatry's Troubled Search for the Biology of Mental Illness* (New York: W. W. Norton, 2019). For more on the clinic, see Gabriel N. Mendes, *Under the Strain of Color: Harlem's Lafargue Clinic and the Promise of an Antiracist Psychiatry* (Ithaca, NY: Cornell University Press, 2015), and Dennis Doyle, "'A Fine New Child': The Lafargue Mental Hygiene Clinic and Harlem's African American Communities, 1946–1958," *Journal of the History of Medicine and Allied Sciences* 64, no. 2 (2009): 173–212.
25. Fredric Wertham, *Seduction of the Innocent* (New York: Rinehart, 1954).
26. Wertham, *Seduction of the Innocent*, 6.
27. Wertham, *Seduction of the Innocent*, 149.
28. John Bowlby, "Forty-Four Juvenile Thieves: Their Characters and Home Lives," *International Journal of Psychoanalysis* 25 (1944): 19–53. See also Michael Follan and Helen Minnis, "Forty-Four Juvenile Thieves Revisited: From Bowlby to Reactive Attachment Disorder," *Child: Care, Health and Development* 36, no. 5 (2010): 639–645.
29. Bowlby, "Forty-Four Juvenile Thieves."
30. Wertham, *Seduction of the Innocent*, 10.
31. Wertham, *Seduction of the Innocent*, 155.
32. Carol L. Tilley, "Seducing the Innocent: Fredric Wertham and the Falsifications That Helped Condemn Comics," *Information & Culture: A Journal of History* 47, no. 4 (2012): 383–413.
33. Lauretta Bender, "The Autistic Child," September 23, 1960, Box 10, Folder 3, the Papers of Dr. Lauretta Bender, Brooklyn College Archives and Special Collections.
34. Bender, "The Autistic Child."
35. Lauretta Bender and Reginald Lourie, "The Effect of Comic Books on the Ideology of Children," presented at the 1941 meeting of the American Psychiatric Association.
36. Lauretta Bender, "The Psychology of Children's Reading and the Comics," *Journal of Educational Sociology* 18, no. 4 (1944): 223–231.
37. Transcript of Bender testimony, Juvenile Delinquency (Comic Books): Hearings before the Subcommittee to Investigate Juvenile Delinquency of the Committee on the Judiciary, United States Senate, Eighty-third Congress, second session, pursuant to S. 190. April 21, 22, and June 4, 1954, last accessed March 10, 2023, https://www.thecomicbooks.com/bender.html.

38. Transcript of Bender testimony.
39. Transcript of Bender testimony.
40. Wertham, *Seduction of the Innocent*, 351.
41. Robert Gorman, "How Much TV Should Children See?," *Redbook*, April 1957.
42. Gorman, "How Much TV Should Children See?"
43. Transcript of Bender testimony.
44. Lynn Spigel, *Make Room for TV: Television and the Family Ideal in Postwar America* (Chicago: University of Chicago Press, 1992), 45.
45. Wertham, *Seduction of the Innocent*, 355.
46. See Meredith A. Bak, *Playful Visions: Optical Toys and the Emergence of Children's Media Culture* (Cambridge, MA: MIT Press, 2020).
47. Wertham, *Seduction of the Innocent*, 51.
48. Elana Levine, *Her Stories. Daytime Soap Opera and US Television History* (Durham, NC: Duke University Press 2020).
49. Spigel, *Make Room for TV*, 193–194.
50. Marshall McLuhan, *Understanding Media: The Extensions of Man* (Berkeley, CA: Gingko, 2013), 212.
51. In the Great Depression, children went home from school with a key around their neck, or one waiting for them, concealed at home. These were the first latchkey children. World War II brought another startling dynamic to the two-parent middle-class family in the United States and Canada (and a different, if not related dynamic in much of Europe): with father away at war, mother frequently filled his economic role and was away at work. Children then went from home in the morning with a single parent present to school, and back home—alone. The term for this child, "latchkey kid," was born. By Joshua Harris's moment in the 1960s, latchkey children were both classed and raced. They were either children of single mothers, needing to take care of themselves and perhaps siblings, or seen as the unfortunate children of working mothers who worked for their own (feminist) fulfillment. Television was called in as adjunctive care, as were libraries, which functioned to fill the gap between the end of the school day and the end of the work day.
52. Ondi Timoner, dir., *We Live in Public*, 2009.
53. Timoner, *We Live in Public*.
54. Josh Harris's experiments with the medium were not just on the increasingly ubiquitous computer screen, but also embodied and offline. As Josh's success grew, he became, in his own words, freer about appearing as his alter ego, Luvvy. When Josh was approximately four years old, he began, like many Americans, to watch *Gilligan's Island*. Broadcast for just three years, the show continues to have iconic status as central to mid-century white domesticity, going off-air timed with the Summer of Love, when that cult momentarily was exchanged

for counterculture. For Josh, no character was more iconic than Eunice Wentworth Howell, better known as Lovey. Lovey became so important to Josh that he created, or became, the character of Luvvy, named for Mrs. Howell. Josh would appear as Luvvy in the boardroom, at Pseudo-sponsored parties in the late 1990s. Those who knew Josh more intimately have said that his costume was not just a version of Lovey, but took elements of costuming and performance from his own mother. His television mother-ideal, Mrs. Howell, his biological mother, and he converged in a new identity—an alter-super ego.

The apotheosis of Harris's work is his most famous experiment, *Quiet: We Live in Public*. In honor of, and leading up to, Y2K, he invited sixty artists to move into a three-floor neo-commune featuring sleep pods, a banquet table, and a gun range—all under surveillance, all being taped as well as subject to "interrogations" and interviews whenever Harris wished. Needless to say, many felt like they were being tortured in a pleasure dome.

55. Bertolt Brecht, "The Radio as an Apparatus of Communication," in *Brecht on Theatre: The Development of an Aesthetic*, ed. and trans. John Willett (New York: Hill and Wang, 1964; originally published as "Der Rundfunk als Kommunikationsapparat" in 1932).

56. Paddy Scannell, "For-Anyone-as-Someone Structures," *Media, Culture, and Society* 22, no. 1 (2000).

57. Linda Charlton, "Surgeon General Wants TV to Curb Violence for Children," *New York Times*, March 22, 1972, https://www.nytimes.com/1972/03/22/archives/surgeon-general-wants-tv-to-curb-violence-for-children.html?searchResultPosition=1.

58. David Buckingham, *Moving Images: Understanding Children's Emotional Responses to Television* (Manchester: Manchester University Press, 1996).

59. Jeanette Steemers, *Creating Preschool Television: A Story of Commerce, Creativity, and Curriculum* (London: Palgrave Macmillan, 2010), 21.

60. Michael Davis, *Street Gang: The Complete History of Sesame Street* (New York: Penguin Books, 2008), 8.

61. Davis, *Street Gang*, 141.

62. Lynn Spigel, *TV Snapshots: An Archive of Everyday Life* (Durham, NC: Duke University Press, 2022).

63. Davis, *Street Gang*, 141–142.

64. C. M. Pierce, "Psychiatric Problems of the Black Minority," in *American Handbook of Psychiatry*, ed. S. Arieti (New York: Basic Books, 1974).

65. Spock as quoted in Davis, *Street Gang*, 198.

66. Les Brown, "Sesame Street Dazzles in Television Premiere," *Variety*, December 24, 1969.

67. Edward Zigler and Susan Muenchow, *Head Start: The Inside Story of America's Most Successful Educational Experiment* (New York: Basic Books, 1994).

68. Quoted in Davis, *Street Gang*, 200.

69. Quoted in Davis, *Street Gang*, 201.

70. "Television: Who's Afraid of Big, Bad TV?," *Time*, November 23, 1970.

71. Robert W. Morrow, *"Sesame Street" and the Reform of Children's Television* (Baltimore: Johns Hopkins University Press, 2005), 47.

72. Arnold Arnold, "Writer Attacks Sesame Street," *Hackensack Record*, August 5, 1970.

73. Linda Francke, "The Games People Play on Sesame Street," *New York Magazine,* April 1971.

74. Francke, "The Games People Play."

75. Morrow, *"Sesame Street" and the Reform of Children's Television*, 147.

76. Stephanie Harrington, "Does Television Hurt the Head? Review of *The Plug-In Drug*," *New York Times*, March 20, 1977. See also C. Lehmann-Haupt, "A Hole into Hell?," *New York Times*, March 7, 1977, 23.

77. Clifford Nass and Brian Reeves, *The Media Equation: How People Treat Computers, Television, and New Media like Real People and Places* (Washington, DC: Center for the Study of Language and Information, 1996).

CHAPTER 6

1. See Elizabeth Patton, *Easy Living: The Rise of the Home Office* (New Brunswick, NJ: Rutgers University Press, 2020).

2. Lynn Spigel, "Designing the Smart House: Posthuman Domesticity and Conspicuous Production," *European Journal of Cultural Studies* (November 1, 2005): 419–420. See also Gwendolyn Brooks, *Building the Dream: A Social History of Housing in America* (New York: Pantheon, 1981).

3. "Abolishing the Family and Obtaining Gestational Justice: An Interview with Sophie Lewis," The Scroll, December 14, 2020, https://politikapolitika.com/2020/12/14/abolishing-the-family-and-obtaining-gestational-justice-an-interview-with-sophie-lewis.

4. Federici writes that women's unpaid labor is to be widely understood "as natural resource or a personal service, while profiting from the wageless conclusion of the labor involved." Silvia Federici, *Caliban and the Witch: Women, the Body and Primitive Accumulation* (Brooklyn, NY: Autonomedia, 2004), 4. For more on the history of domestic labor and economics in the US context vis-à-vis gender and family work, see Dolores Hayden, *Redesigning the American Dream: Gender, Housing, and Family Life* (New York: W. W. Norton, 1984), 121–140. For a synthetic reading of the "Wages for Housework Movement" and its preconception in the present, see Kathi Weeks, *The Problem with Work: Feminism, Marxism, Antiwork Politics, and Postwork Imaginaries* (Durham, NC: Duke University Press, 2011).

5. In order to situate "the political and temporal narratives of humanization," Neda Atanasoski and Kalindi Vora use the concept of the "surrogate human" as "one way to conceptualize the function of the technologized posthuman stand-in, a rich and experimental subject/object encompassing the techno-fantasies . . . through which racialized, gendered, and sexualized

spheres of life and labor are seemingly elided by technological surrogates, even as these spheres are replicated in emergent modes of work, violence, and economies of desire." N. Atanasoski and K. Vora, "Surrogate Humanity: Posthuman Networks and the (Racialized) Obsolescence of Labor," *Catalyst: Feminism, Theory, Technoscience* 1, no. 1 (2015): 2–3.

6. Hayden, *Redesigning the American Dream*, 93–96.

7. For a brief history of the lag between the electric grid and the wiring of individual homes, see Jason E. Smith, *Smart Machines and Service Work: Automation in the Age of Stagnation* (Chicago: University of Chicago Press, 2020).

8. "'Home Electrical' Wins Award at World's Fair," *Sedalia Democrat*, November 14, 1915, 16.

9. For more on the provenance of this film and its history, see Martin L. Johnson, "Establishing the Provenance of Early Advertising Films: Film Catalogs and the Creation of the Nontheatrical Market," in *Provenance and Early Cinema*, ed. Joanne Bernardi, Paolo Usai, Tami Williams, and Joshua Yumibe (Bloomington: Indiana University Press, 2021), 214–222.

10. Unfortunately, *Every Husband's Opportunity* has been lost. This description comes from "Films to Popularize Electric Services," *Motography*, June 1912, 245.

11. John W. Schwem, "My 45 Years of Business Film Making," *Business Screen Magazine* 2, no. 3 (1954): 68.

12. "Films to Popularize Electric Services," *Motography*, June 1912, 245. For more on the use of film in advertising technology in this period and subsequently, see Joseph Corn, "Selling Technology: Advertising Films and the American Corporation 1900–1920," *Film & History* 11, no. 3 (September 1981): 49–58.

13. "Films to Popularize Electric Services," 245. For more on the domestic working conditions of both housewife and nursemaid, see Susan Strasser, *Never Done: A History of American Housework* (New York: Owl Books, 1982).

14. "The Home Electrical," General Electric, 1915, Schenectady Innovation and Technology Museum.

15. Sianne Ngai, *Theory of the Gimmick* (Cambridge, MA: Harvard University Press, 2020).

16. Claudia Goldin, "The Work and Wages of Single Women, 1870–1920," *Journal of Economic History* 40, no. 1 (1980): 81–88.

17. Hayden, *Redesigning the American Dream*, 59.

18. Lynn Spigel, *Welcome to the Dreamhouse: Popular Media and Postwar Suburbs* (Durham, NC: Duke University Press, 2001), 404. For more on the aesthetics of nostalgia and progress in suburbia, see Holley Wlodarczyk, "This Old House of the Future: Remixing Progress and Nostalgia in Suburban Domestic Design," in *Marking Suburbia: New Histories of Everyday America*, ed. John Archer, Paul J. P. Sandul, and Katherine Solomonson (Minneapolis: University of Minnesota Press, 2015), 164–183.

19. Stephanie Coontz, *The Way We Never Were: American Families and the Nostalgia Trap* (New York: Basic Books, 2016).

20. Lisa D. Schrenk, *Building a Century of Progress: The Architecture of the Chicago's 1933–1934 World's Fair* (Minneapolis: University of Minnesota Press, 2007), 138–139. For more on the history of model homes in Chicago, see Gwendolyn Brooks, *Moralism and the Model Home: Domestic Architecture and Cultural Conflict in Chicago, 1873–1913* (Chicago: University of Chicago Press, 1980).

21. Schrenk, *Building a Century of Progress*, 139.

22. Schrenk, *Building a Century of Progress*.

23. For a history of "creative" children's rooms in this period, see Amy Ogata, *Designing the Creative Child: Playthings and Places in Midcentury America* (Minneapolis: University of Minnesota Press, 2013). For more on design movements centering children and their play, see Juliet Kinchin and Aidan O'Connor, *Century of the Child: Growing by Design, 1900–2000* (New York: MoMA, 2012).

24. Karin Calvert, *Children in the House: The Material Culture of Early Childhood, 1600–1900* (Boston: Northeastern University Press, 1994), 133.

25. Noguchi, letter to E. F. McDonald Jr., 1946, MS_PROJ_057_004, Manuscript Collection, Project Files, Noguchi Archives.

26. Patton, *Easy Living*.

27. David Snyder, "Playroom," in *Cold War Hothouses: Inventing Postwar Culture, from Cockpit to Playboy,* ed. Beatriz Colomina, Ann Marie Brennan, and Jeannie Kim (Princeton, NJ: Princeton Architectural Press, 2004), 129. This is perhaps why Bradbury is able to describe a virtual reality playroom, but instead *calls* it a nursery—it is both commonplace and a new phenomenon in his moment.

28. For more on homes as they were lived in and built in this moment, see, as just a few examples: Keller Easterling on the split-level house in *Organization Space: Landscapes, Highways, and Houses in America* (Cambridge, MA: MIT Press, 1999); Thomas C. Hubka, *How the Working-Class Home Became Modern, 1900–1940* (Minneapolis: University of Minnesota Press, 2020); Dianne Harris, *Little White Houses: How the Postwar Home Constructed Race in America* (Minneapolis: University of Minnesota Press, 2013); Gwendolyn Brooks, *Building the Dream: A Social History of Housing in America* (New York: Pantheon, 1981); James A. Jacobs, *Detached America: Building Houses in Postwar Suburbia* (Charlottesville: University of Virginia Press, 2015); Barbara Miller Lane, *Houses for a New World: Builders and Buyers in American Suburbs 1945–1965* (Princeton, NJ: Princeton University Press, 2015); Andrew Shanken, *194X: Architecture, Planning, and Consumer Culture on the Home Front* (Minneapolis: University of Minnesota Press, 2009).

29. For an extensive look at the long-term effects of redlining practices in the United States, see "The House We Live in," Race: Power of an Illusion, Episode Three, California Newsreel, 2003.

30. Spigel, *Welcome to the Dreamhouse*, 110.

31. Spigel, *Welcome to the Dreamhouse*. For more on the history of the prefab home and its marketing, see Anna Vemer Andrezejewski, "Selling Suburbia: Marshall Erdman's Marketing Strategies for Prefabricated Buildings in the Postwar United States," in *Marking Suburbia: New Histories of Everyday America*, ed. John Archer, Paul J. P. Sandul, and Katherine Solomonson (Minneapolis: University of Minnesota Press, 2015), 281–304.

32. "Fueled by Aging Baby Boomers, Nation's Older Population to Nearly Double in the Next 20 Years, Census Bureau Reports," United States Census Bureau, May 6, 2014.

33. This is far from an exhaustive account of all "smart homes" in the United States, nor is it a total account of futuristic architecture. There are many smart homes in this lineage, including both model homes and homes that were actively lived in. For another set of smart homes to consider, see Spigel, "Designing the Smart House."

34. For more on the Cold War politics of architectural domestic design in this moment, see Greg Castillo, *Cold War on the Home Front: The Soft Power of Midcentury Design* (Minneapolis: University of Minnesota Press, 2010).

35. For a critical history of Disney theme parks, see Sabrina Mittermeier, *Middle Class Kingdoms: A Cultural History of Disneyland Theme Parks* (London: Intellect Press, 2020).

36. "Monsanto House of the Future," Monsanto, 1957, YouTube, last accessed December 1, 2020, https://www.youtube.com/watch?v=sk2YBA_oa1A

37. "Monsanto House of the Future."

38. "Monsanto House of the Future." For more on the valences of Monsanto and the use of plexiglass, see Shannon Mattern, "Towards a Cultural History of Plexiglass," *Places Journal*, December 2020, https://placesjournal.org/article/purity-and-security-a-cultural-history-of-plexiglass/.

39. For more on the American dream house, and the suburban homes of this moment, see "Nationalizing the Dream" in John Archer, *Architecture and Suburbia: From English Villa to American Dream House, 1690–2000* (Minneapolis: University of Minnesota Press, 2005).

40. Snyder, "Playroom," 140.

41. Snyder, "Playroom," 73.

42. Ogata, *Designing the Creative Child*, 73.

43. Jordynn Jack, *Autism and Gender: From Refrigerator Mothers to Computer Geeks* (Champaign: University of Illinois Press, 2004), 45.

44. The House of the Future was an exercise in nostalgia for a lost future, or futural nostalgia. Lynn Spigel notes that these terms are not antonymic, but two strategic displacements of the present. The house, neither a sci-fi caricature nor an educated guess as to how one might live in the future, assumed that *everything* outside the home and all the people in it would remain unchanged. Spigel, *Welcome to the Dreamhouse*, 404. For more on the aesthetics of nostalgia and progress in suburbia, see Wlodarczyk "This Old House of the Future," 164–183. For more of Monsanto's work at the New York City World's Fair, see *Monsanto Magazine*, Summer

1964, last accessed December 1, 2020, http://www.worldsfairphotos.com/nywf64/articles/misc/monsanto-magazine-summer-1964.pdf.

45. Ruth Schwartz Cowan, *More Work for Mother: The Ironies of Household Technology from the Open Hearth to the Microwave* (New York: Basic Books, 1985), 195; Judy Wajcman, *Feminism Confronts Technology* (Cambridge: Polity Press, 1991), 44–109.

46. Cowan, *More Work for Mother*, 195.

47. "A Day in the Life of a Kitchen," Frigidaire Division, General Motors Corp., 1965, Prelinger Archives, https://archive.org/details/Day_in_the_Life_of_a_Kitchen_A; *It Happened in the Kitchen* (General Electric, 1941), YouTube, https://www.youtube.com/watch?v=ndhcaVMawrI; *A Step-Saving Kitchen* (United States Department of Agriculture 1949), the National Archives, Washington, DC, YouTube, https://www.youtube.com/watch?v=2N9RCQjPqh4.

48. Westinghouse Leisure Appliances Advertisement, c. 1940. In Ellen Lupton, *Mechanical Brides: Women and Machines from Home to Office* (Princeton, NJ: Princeton Architectural Press, 1993), 14.

49. For more on the importation of factory management into the home, see Arlie Hochschild, *The Commercialization of Intimate Life* (Berkeley: University of California Press, 2003).

50. Mar Hicks, "Sexism Is a Feature, Not a Bug," in *Your Computer Is on Fire*, ed. Thomas S. Mullaney, Benjamin Peters, Mar Hicks, and Kavita Philip (Cambridge, MA: MIT Press, 2021).

51. Joy Lisi Rankin, *A People's History of Computing in the United States* (Cambridge, MA: Harvard University Press, 2018), 125–126.

52. Roy Mason with Lane Jennings and Robert Evans, *Xanadu: The Computerized Home of Tomorrow and How It Can Be Yours Today!* (Washington, DC: Acropolis Books, 1983), 19.

53. Hicks, "Sexism Is a Feature, Not a Bug."

54. Mitra Toossi and Teresa L. Morisi, "Women in the Workforce before, during, and after the Great Depression," Bureau of Labor Statistics, last accessed July 14, 2020, https://www.bls.gov/spotlight/2017/women-in-the-workforce-before-during-and-after-the-great-recession/pdf/women-in-the-workforce-before-during-and-after-the-great-recession.pdf.

55. Mason et al., *Xanadu*, 1.

56. Mason et al., *Xanadu*, 44–45.

57. Mason et al., *Xanadu*, 119.

58. Mason et al., *Xanadu*, 84. For more on other robotic smart home assistants, see Spigel, "Designing the Smart House"; Thao Phan, "Amazon Echo and the Aesthetics of Whiteness," *Catalyst: Feminism, Theory, and Technoscience* 5, no. 1 (2017); Yolande Strengers and Jenny Kennedy, *The Smart Wife: Why Siri, Alexa, and Other Smart Home Devices Need a Feminist Reboot* (Cambridge, MA: MIT Press, 2020); and Kylie Jarrett, *Feminism, Labor, and Digital Media: The Digital Housewife* (London: Routledge, 2017).

59. Mason et al., *Xanadu*, 84.
60. Mason et al., *Xanadu*, 19.
61. Laine Nooney, "'The Computerized Home of Tomorrow': The Xanadu Homes and the American Fantasy of Privatized, Computational Living," March 16, 2018, Society for Cinema and Media Studies, Toronto, Canada.
62. For more on this, see Arlie Hochschild, "The Commodity Frontier," in *The Commercialization of Intimate Life: Notes from Home and Work* (Berkeley: University of California Press, 2003).
63. For more on postwar car culture as an extension of domesticity, see Beatriz Colomina, "Cold War/Hothouses," in *Cold War Hothouses: Inventing Postwar Culture, from Cockpit to Playboy*, ed. Beatriz Colomina, Ann Marie Brennan, and Jeannie Kim (Princeton, NJ: Princeton Architectural Press, 2004), 12–14.
64. Cowan, *More Work for Mother*, 189.
65. For a longer history of work from home, the home office, and gender, see Patton, *Easy Living*.
66. Spigel, "Designing the Smart House," 419–420. For an extended consideration of work from home and its impacts, see Ursula Huws, *Teleworking and Gender* (Brighton: Institute for Employment Studies, 1996); Melissa Gregg, *Work's Intimacy* (London: Polity Press, 2011). For an overview of women's work in this period, see Lynn Y. Weiner, *From Working Girl to Working Mother: The Female Labor Force in the United States, 1820–1980* (Chapel Hill: University of North Carolina Press, 1985), 114–140; for more on feminized work and digital media, see Brooke Erin Duffy, *(Not) Getting Paid to Do What You Love* (New Haven, CT: Yale University Press, 2017).
67. For more on the valances of flexibility in telework and the "electric cottage," see Patton, *Easy Living*, 145–155.
68. Patton, *Easy Living*.
69. Mason et al., *Xanadu*, 151.
70. Mason et al., *Xanadu*, 152.
71. David Rakoff, "The Future Knocks Again," *New York Times*, July 10, 2008.
72. Rakoff, "The Future Knocks Again." Disney also released a film, "Smart House," the following year, after this model home debuted. It follows a plot nearly identical to Bradbury's story. A young teen wins the Smart House in a giveaway after the death of his mother. The home, having replaced mother, takes on a life of its own and controls the family, as an "overbearing mother." LeVar Burton, dir., *Smart House* (Disney Channel, 1999).
73. For more on the modes in which this panic operates, as well as resistance to it, in the late twentieth and early twenty-first century, see Grace Chang, *Disposable Domestics: Immigrant Workers in the Global Economy* (Chicago: Haymarket Books, 2016).
74. And while iPads are attached to cribs and handed to toddlers so that parents can get dinner on the table, answer work email, or attend to other things, the devices do not directly remediate

the nanny or the mother herself, but instead are an iteration of media for pacification, a tool at the disposal of mothers and caregivers since the mid-1800s. For more on the early history of children's media for pacification and play, see Meredith Bak, *Playful Visions: Optical Toys and the Emergence of Children's Media Culture* (Cambridge, MA: MIT Press, 2020); for mid-century cultures, see Spigel, *Make Room for TV*.

75. See Natasha D. Schüll, "The Sense Mother," *Fieldsights* (October 31, 2018). See also the work of Sarah Sharma, who links the devaluation of maternal care labor with the devaluation of gig economy and labor platforms in Sarah Sharma, "Task Rabbit: The Gig Economy and Finding Time to Care Less," in *Appified: Culture in the Age of Apps*, ed. Jeremy Wade Morris and Sarah Murray (Ann Arbor: University of Michigan Press, 2018), 63–71.

76. Arlie Hochschild, *The Second Shift: Working Families and the Revolution at Home* (New York: Penguin Books, 2012).

77. Hochschild, *The Second Shift*.

CODA

1. For more biographical information on Rudolf Steiner, see Gary Lachman, *Rudolf Steiner* (New York: Penguin, 2007).

2. Peter Staudenmaier, *Between Occultism and Nazism: Anthropology and the Politics of Race in the Fascist Era* (Leiden: Brill, 2014).

3. P. Bruce Uhrmacher, "Uncommon Schooling: A Historical Look at Rudolf Steiner, Anthroposophy, and Waldorf Education," *Curriculum Inquiry* 25, no. 4 (1995): 381–406.

4. Rudolf Steiner, "Spiritual Knowledge as a Way of Life," lecture at The Hague, November 1923. Trans. Mary Adams, in *The Golden Blade*, 1950 [first English translation].

5. Rudolf Steiner, *On Epidemics: Spiritual Perspectives* (Forest Row: Rudolf Steiner Press, 2012), 20.

6. Peter Staudenmaier, "Race and Redemption: Racial and Ethnic Evolution in Rudolf Steiner's Anthroposophy," *Nova Religio* 11, no. 3 (February 2008): 4–36.

7. Staudenmaier, *Between Occultism and Nazism*.

8. Kimiko de Freytas-Tamura, "Bastion of Anti-Vaccine Fervor: Progressive Waldorf Schools," *New York Times*, June 13, 2019; and Rachel Monahan, "Ending the Pandemic Means Getting Vaccinated. But Many Oregonians Will Be Hard to Convince," *Willamette Weekly*, January 6, 2021.

9. de Freytas-Tamura, "Bastion of Anti-Vaccine Fervor."

10. See, for instance, the Steiner Fellowship's (UK and Ireland) message to affiliate schools, Fran Russell, "Covid Vaccinations: Advice to Schools," February 4, 2021, last accessed in May 2021, https://www.steinerwaldorf.org/covid-vaccinations-advice-to-schools [discontinued].

11. Morgan G. Ames, "The Smartest People in the Room? What Silicon Valley's Supposed Obsession with Tech-Free Private Schools Really Tells Us," *Los Angeles Review of Books*, October 18, 2019.

12. Danya Glabau, "Necessary Purity," *Real Life*, June 6, 2017.

13. Nancy McGrath, "Learning with the Heart," *New York Times*, September 25, 1977.

14. Otto Specht School, internal communication to families, "The Immune Sequence," September 23, 2020.

15. Samuel Weber, *Preexisting Conditions: Recounting the Plague* (Princeton, NJ: Zone Books, 2022).

16. Laura Portwood Stacer, "How We Talk about Media Refusal, Part 2: Asceticism," *Flow Journal*, September 10, 2012. For more on the internal protocols of MOVE, see Richard Kent Evans, *MOVE: An American Religion* (Oxford: Oxford University Press, 2020).

Index

Note: Page numbers in italics indicate illustrations.

Achievement feminism, 108
Active play therapy, 126
Adoption, 102, 126, 153
Adoption and Safe Families Act (1997), 166
Adorno, Theodor, vii, 22
Affection. *See* Attention and affection
African Americans. *See also* Captive Maternals
   child psychology and, 187–188, 193–194
   and children's status, 49
   family dynamics of, 130–131
   as hot mothers, 130–131, 134, 269n42
   natal alienation of, 27
   schizophrenia linked to, 133–134
   as television audience, 206–207, 209
   violence experienced by, 190, 192, 206–207
   in Wisconsin criminal justice system, 145–147
Ainsworth, Mary, 142, 152, 161–162, 168, 171, 271n82
Air Crib, 91–92, *93*, 94–99, *96*, 106, 263n36, 264n49
Alexa (virtual assistant), 230
Allami, Jason, 71–72
Allen, Woody, *Zelig*, 136
Always-on mothers, 5, 29, 143, 218, 235
American Psychiatric Association, 193
American Psychoanalytic Association, 18, 126

Ames, Morgan, 241, 242
Anthroposophy, 237–241
Anti-Drug Abuse Act (1986), 164
Anti-mediation, 7
Anzieu, Didier, 15
Apple, Rima D., 11
Apple I, 106
Appliances, 65, 134–138, 141, 218–219, 222–223, 226–227. *See also* Television
Arnold, Arnold, 209
Artificial mothering, 102
Asians/Asian Americans, 142
Asperger, Hans, 269n50
A.S. v. N.S. (2020), 168
Atanasoski, Neda, 39, 284n5
Attachment, as criterion of maternal fitness, 157
Attachment parenting. *See also* Attachment theory
   attachment theory in relation to, 104, 166, 271n82
   development of, 103–105
Attachment theory. *See also* Attachment parenting
   attachment parenting in relation to, 104, 166, 271n82
   basis of, 142, 152

Attachment theory (cont.)
  and day nurseries, 63, 153
  and delinquency, 191
  development of, 151–152
  documentary component of, 161, 168
  helicopter parenting and, 142
  historical context for, 159–160
  legal uses of, 163, 166–168
  and mother's role, 15, 114, 119–120
  parallel trajectories of, 163–164, 167–168
  prisons as site for studying/applying, 149–150, 165–172, 279n93
Attention and affection
  expectations placed on mothers concerning, 202–203
  harms from deprivation of, 94, 113–115, 118–121, 126, 131–140
  harms imputed to, 84–85, 88, 114–115, 121–131
Autism, 113, 114–115, 117, 121, 124, 131–134, 136–137, 141, 193, 267n17, 270n65
Automata. *See* Robots and automata

Babies Hospital, Newark, New Jersey, 44
Babies Hospital, New York, 84
Baby boom, 222–223
Baby jumpers, 87
Baby M, 102
Baby monitors, 60–62, *61*, 65, 108–109. *See also* Nanny cams
Babysitter panics, 63
"Back to Sleep" campaign, 107
Baldwin, James Mark, 10
Baraitser, Lisa, 26
Barthes, Roland, 79–80
Bassinets, 102
Bateson, Gregory, 17, 125–126
Bedford Hills Correctional Facility, Westchester County, New York, 169–171, *170*, 173
Bedwetting, 90

Beecher, Catherine, 48
Behar, Yves, 77
Behaviorism, 11, 53–54, 89–95, 114. *See also* Neo-behaviorism
Bellevue Psychiatric Hospital, New York, 12, 17, 134, 187, 192–194
Bender, Lauretta, 17, 22, 133–134, 192–195, 197
Benjamin, Jessica, 18
Bernstein, Robin, 49
Bettelheim, Bruno, 134, 136–140, 203, 275n36
Bill 627 (Wisconsin), 145, 147, 163, 166
Bion, Wilfred, 15, 105
Black, Irma Simonton, 196
Black Psychiatrists of America, 206
Boor, Clemens de, 253n42
Booth, William E., 165
Boston Process Group, 18
Bowlby, John, 15, 119–120, 142, 147, 152–154, 191, 271n82
Boyer, Anne, 43
Boys, advice for raising, *123*, 127, 129
Bradbury, Ray, "The World the Children Made" (now called "The Veldt"), *x*, 1–6, 20–21, 66, 225–226, 228, 230, 232, 234, 289n72
Brady, Mary Pat, 49, 146
Breastfeeding/breast milk, 45–46, 64, 81, 84, 103–104
Brecht, Bertolt, 201
Bronfenbrenner, Urie, 208
Brown, Les, 208
Browne, Simone, 74
Brown v. Board of Education (1954), 188
Buckingham, David, 204
Buxton, William J., 184
Byrne, Mary, 171

Cairns, Robert B., 119–120
Calvert, Karin, 47–48, 93
Campbell, Nancy, 158

Canguilhem, Georges, 16
Cantril, Hadley, 185
Capitalism, 49, 79, 143, 146. *See also* Carceral capitalism
Capone, Al, 36
Captive Maternals, 148, 158, 164, 172–173
Carceral capitalism, 146–147, 170, 173
Carnegie Foundation, 205
Cartwright, Lisa, 156, 161, 275n36
Catalan, Ilse Denisse, 158
Cavett, Dick, 136
CBS (TV network), 198
Charcot, Jean-Martin, 12–13
Chicago Daily News (newspaper), 186
Chicago School of sociology, 24, 182, 268n40
Childcare. *See also* Mothers/mothering; Nannies/nurses; Parenting/child-rearing
anxiety about, 29, 68–69, 72, 234
crisis (problems and barriers) of, 25, 29, 68, 70, 135, 216, 228, 234–235
mothers' work conducted simultaneously with, 41–42, 232
paid labor for, 43
in prisons, 149–151, 153–160, 163–173
race and, 28
smart homes/suburban homes and, 217, 221–222, 225, 227–229, 232, 235
Child development theory/field. *See also* Parenting theory; Psychology
and the Air Crib, 97
formation of, 10
nature vs. nurture in, 10–11, 14
in policy and popular contexts, 16–17
professionalization of, 12
study of media by, 180
study of mothers/mothering by, 9, 11
study of the family by, 9
Child Protective Services (CPS), 27, 145, 147–148, 164
Child psychiatry, 11, 13–14, 53–54, 121–122
Child psychoanalysis, 11, 13–14, 22, 55, 151. *See also* Attachment theory

Child-rearing. *See* Mothers/mothering; Nannies/nurses; Parenting/child-rearing
Children. *See also* Containment, of children; Maternal deprivation; Pacification; Scientific studies of children
characterological threats in upbringing of, 11, 84, 88, 92–93
comic books' effect on, 17, 22, 186–195
contamination threats to, 46
"emotionally disturbed," 134, 136–137, 193–194
impressionability of, 186
innocence imputed to, 49
meanings attached to, 49
movies' effect on, 181–186
natal alienation of, 27
outside influences on, 46–48
as priceless, 49, 67, 81
purity imputed to, 241–243
rooms for, 1–2, 64, 221–222, 224–225, 232, *233*
surveillance of, 40, 63–66, *66*, 74
television's effect on, 195–213
threats to, 11, 70
training of, 89
Children's Welfare Bureau, 9, 42
Children's Workshop, 206
Child Study Association of America, 22
Child Welfare Bureau, 148. *See also* Child Protective Services
Chodorow, Nancy, 18, 140
Chun, Wendy, 67, 274n23
Cinema. *See* Movies
Class
children's status and, 49–50
incarceration practices and, 145–147
marital roles and, 235
mothering styles and, 118
nannies and, 59–60, 74–75
and sleep practices, 83
smart homes and, 235
Clayton, Angelica, 19

Clicker training, 127
Clinton, Bill, 164, 166
Closed-circuit television, 65, *66*, 69, 223. *See also* Nanny cams
Coding, of behaviors, 161–163
Cold War, 26, 100, 115, 126, 129, 149, 159, 185, 223
Collier's Magazine, 187, 189
Colman, Mildred, 177
Colombo, Daria, 52
Columbia University Psychoanalytic Clinic, 126
Comic book panics, 186–195
Comic books, 17, 22, 186–195, 198
Communication, mother-child, 125–126
Computers in the home, 227–229, 231
Comte, Auguste, 16
Conditioning. *See* Behaviorism
Conference on Homebuilding and Homeownership, 54
Connolly, Brian, 37–38
Contagion. *See* Contamination; Illness/contagion; Media contagion; Moral contagion; Social contagion
Containment, of children, 8, 15, 19, 28, 48, 63–64, 80, 82, 83, 87, 89, 97, 99, 103, 104, 113, 150, 202, 221
Contamination
 of the family, 44, 48
 mother as source of, 9
 nannies/wet nurses as source of, 45–47, 62
Continuum concept, 103, 271n82
Cook, Thomas D., 210
Cool mothers, 15, 113–115, 130–140
Cooney, Joan Ganz, 205
Coontz, Stephanie, 3, 28, 38, 221
Cooper, Melinda, 31
Correctional Offender Management Profiling for Alternative Sanctions (COMPAS), 146
Co-sleeping, 82–84, 104, 106
Couch potatoes, 180

Council on Mental Health of the American Medical Association, 197
COVID-19, 29, 30, 216, 240–243
Cowan, Ruth Schwartz, 53, 54, 65, 109, 135, 218, 220, 227, 231
CPS. *See* Child Protective Services
Cradles, 48, 84, 86–88, 89, 262n20
Cressey, Paul, 184–185
Cribs, 28, 48, 64, 79, 84, 88, 89, 93, 95, 103–107, 222. *See also* Air Crib
Crown Heights, Brooklyn, New York, 177
Cult of Domesticity, 27
Cult of True Womanhood, 27
Cybernetics, 13, 17, 19, 97, 128, 161, 274n23

Darwin, Charles, 80
Davis, Angela, 28
Davis, Michael, 205, 209
"Day in the Life of a Kitchen, A" (TV short), 227
Day nurseries. *See* Nursery schools/day nurseries
Dead mother complex, 140
Delinquency
 Bowlby and, 152, 191
 comic books linked to, 186, 190–191
 mid-twentieth-century concern with, 149, 152, 159–160, 190, 194
 mothers' role in, 23, 130–131, 135, 159–160, 172
 movies linked to, 183
 psychosocial causes of, 189, 191–192
 Senate hearings on, 194–195
Democracy, mothering's influence on, 26, 127–128
Demos, John, 7
Deprivation. *See* Maternal deprivation
Detachment parenting, 103
Dewey, John, 10
Disney, 233, 234, 289n72. *See also* Walt Disney Imagineering

Disneyland, 223
Disney World, 230
Doane, Mary Ann, 156
Dolls, play therapy involving, 126
Domestic workers. *See also* Nannies/nurses
  decline of, in interwar period, 38, 53–55
  media/technology as stand-ins for, 39, 65
Dorshow, Deborah Blyth, 90
Double bind
  defined, 125
  of mothering, 20, 29, 57, 60, 68, 75, 108, 125, 139, 144
  of nannying, 59
  of time/labor-saving devices, 218, 220
  of women's sexuality, 129
  women's work-home, 108, 135, 215–216
Dream Home, 233–234
Duess test, 190
Duong, Kevin, 187–188

Eappen family, 71
Education and schools, 184, 205–211, 237–243
Ego psychology, 13
Eisenberg, Leon, 208
Electric cradles, 86–88, *87*, 262n20
Electricity, 86, 137–139, 218–220. *See also* Appliances
Elephant Man, 10
Ellison, Ralph, 188
El Wardany, Haytham, 110
"Emotionally disturbed" children, 134, 136–137, 193–194
Engelhardt, Tom, 23
Enslaved persons
  technological substitutes for, 39, 62
  as wet nurses, 10, 45
Environment. *See* Maternal environment
Epcot Center, Disney World, 230
Epstein, Mel, 178
Erikson, Erik, 16, 17, 22, 203
Essanay Motion Picture Company, 219

Ethnopsychoanalysis, 161
Eubanks, Virginia, 274n23
Eugenics, 84, 148, 172, 192
Eurythmy, 238, 242
Every Husband's Opportunity (film), 219

Family. *See also* Multigenerational families; Nuclear family
  as anecdote to mediation, 1–3
  contamination of, 44, 48
  crisis of, 5, 7, 31
  factors contributing to a healthy, 78
  importance and responsibility attributed to, 25
  mid-twentieth-century ideal of, 3–4
  natal alienation and, 27
  outside influences on, 3–4, 21–22, 25–26, 38, 73
  psychological and other social scientific study of, 4–9
  purity of, 7, 28
  self-sufficient and self-reliant ideal of, 3–4, 25–26, 38, 63
  sleep's role in the health of the, 78, 82, 95
  as source of life's gratification, 1, 3
  stability as criterion for judging, 145, 147–148, 157
Fascism, 16, 26, 128, 190, 239, 243
Fashion systems, 79–80
Fass, Paula, 10, 35
Fathers
  Black, 130–131
  in psychological considerations of child-rearing, 14, 55–56, 129
Feder, Luis, 253n42
Federici, Silvia, 217, 284n4
Feedback, 97
Feldstein, Ruth, 130
Feminism
  achievement feminism, 108
  on childcare, 29
  criticisms of, 129

Feminism (cont.)
    and critique of women's home life, 218, 226
    on mothers/mothering, 17–18, 48
    and technology, 217, 255n19
    on work-home double bind, 108
Ferber, Richard, 105–106
Ferenczi, Sándor, 251n20
Films. *See* Movies
Fissell, Mary, 10
Fitness, for motherhood
    attachment as factor in, 157
    criteria of, 148
    incarceration and, 147, 164, 167
    in Progressive era, 276n53
    racial component of, 9, 172
    stability as factor in, 157
    stimulation (excessive or deficient) of children as factor in, 117
Forman-Brunell, Miriam, 63
Formula, baby, 46, 81
Foster care, 145, 166–167
Francke, Linda, 209
Frankfurt school, 22
Frazier, E. Franklin, 268n40
Freedman, Estelle B., 148, 158
Freud, Anna, 13–14, 119, 152
Freud, Sigmund
    American popularity of, 2, 12
    criticisms of, 122–123
    death drive concept of, 251n20
    and family psychology, 38
    and mother-child psychology, 12–14, 55, 130
    and nannies/nurses, 51–53
    parenting advice based on theories of, 123
    and schizophrenia, 137
Friedan, Betty, *The Feminine Mystique*, 137
Frigidaire, 227
Frosted babies. *See* Refrigerator mothers
Fuller, Kathryn H., 182
Furniture, 47–48, 63, 91, 103

Garfunkel, Frank, 208
General Electric, 218–221
Geoghegan, Bernard, 8, 150, 254n61, 274n23
Gerbner, George, 211
Gesell, Arnold, 12, 97, 161
GI Bill, 222
Gilligan's Island (TV show), 282n54
Glabau, Danya, 241
Glover, Edward, 251n20
Goffman, Erving, 25
Goldin, Claudia, 220
Good-enough mothers, 55, 120
Good Housekeeping (magazine), 59, 63–64, 221
Good Morning America (TV show), 102
Goody, Marvin, 223
Goodyear, 221
Googie architecture, 66, 223
Gorman, Robert, 196–197
Governesses. *See* Nannies/nurses
Gow, Betty, 35
Gray, John, Jr., *93*
Great Depression, 42, 49–50, 54, 62, 221, 222, 282n51
Green, André, 140
Green, Emily, 58
Greenacre, Phyllis, 15, 17
Grief: A Peril in Infancy (film), 154–156, *155*, 275n36
Groomer, Dorothy, 134
Gruenberg, Sidonie, 192
Guillory, John, 20, 21
Gumbs, Alexis Pauline, 26

Hall, Martha Haygood, 59
Hamilton, Richard, 223
Happylife Home, 1–2, 225
Haraway, Donna, 30
Harlow, Harry, 21, 100, 119–121, *120*, 128, 267n17
Harris, Josh, 200–201, 282n54

Harris, Malcolm, 143
Hatch, 107
Hauptmann, Bruno, 36
Hayden, Dolores, 255n19
Hays, Sharon, 29
Hays Code, 185–186, 189
Head Start, 205, 208–209
Heartbeat, 100–101
Helicopter parents, 115, 141–144
Henry, Kelsey, 49, 81
Herman, Ellen, 18, 64, 126, 129, 153, 160
Hewlett-Packard, 234
Heyes, Cressida, 104
High chairs, 28, 48, 81, 103
Hochschild, Arlie, 235
Holt, Luther Emmett, 84, 89, 95
   Care and Feeding of Children, The, 84, 105
Home construction and home ownership, 222–223
Home Electrical, 218–221, 233
Home Electrical, The, (film), 219–220
"Home of Tomorrow, The," 221, 233
Homes. *See* Model homes; Smart homes; Suburban homes/suburbia
Home work, 87, 202, 215, 220. *See also* Telework
Homosexuality, 129, 190. *See also* Lesbian mothers
Honeywell Home Kitchen Computer, 227
Hooton, Earnest A., 199
Hoover, Herbert, 36, 54
Horney, Karen, 14, 140
Hot mothers, 114–115, 121–131, 269n42
House of the Future, 223–226, *224*, 230, 233, 287n44
Houses. *See* Home construction and home ownership; Model homes; Smart homes; Suburban homes/suburbia
Housewives
   bored, 115, 226
   ideal/fantasy of, 9, 28, 65, 104
   technological supplements for, 91–92, 106, 135–136, 219, 223–229
   and television, 198
Hu, Tung-Hui, 180
Hug-Hellmuth, Helen, 13
Hughes, Dorothy Pitman, 209
Hygiene, and sleep practices, 82–83
Hypodermic effect, 181
Hysteria, 51

Icebox babies. *See* Refrigerator mothers
Illness/contagion, 237–243
Immersion mothering, 104
Immigrants, natal alienation of, 27
Implicit Association Test, 190
Impulses. *See* Maternal impulses
Incarceration. *See* Prisons/incarceration
Independent sleep, 77
Indigenous peoples, natal alienation of, 27
Infant-mother psychic matrix, 16
Instincts. *See* Maternal instincts
Intensive mothering/parenting, 29, 95, 103, 225–226, 228, 235
International Psychoanalytic Association, 18
iPad, 179, 213, 289n74
Irani, Lily, 73
It Happened in the Kitchen (documentary), 227

Jack, Jordynn, 122, 133
Jackson, Zakiyyah, 49
James, Joy, 148, 158, 172
Jameson, Fredric, 30
Jarvie, Ian C., 182
Jetsons, The (TV show), 66
Joey (case-study subject), 137–139
Johns Hopkins Harriet Lane House, 12, 53–54
Johns Hopkins Phipps Clinic, 13, 16
Johns Hopkins University, 113, 122, 133–134, 193
Johnson, Donald L., 119–120

Johnson, Walter, 148
Joice, Katie, 155, 171
Jolowicz, Ernst, 188
Jones-Rogers, Stephanie E., 45, 46, 130
Joseph, Edward, 253n42
Jowett, Garth S., 182

Kanner, Leo, 13, 17, 53, 113, 117, 121–127, 131–137, 140, 193, 208, 267n17, 269n50
  In Defense of Mothers, 122–123, 131, 136
Kardiner, Abram, 269n42
Karp, Harvey, 77
Karp, Nina Montee, 77
Katz, Elihu, 185
Keck, George Fred, 221
Kefauver hearings, 194–195
Kelly, Robert, 215–216
Key, Elizabeth, 148
Kim, Jung-a, 215
Kinsey, Alfred, *Sexual Behavior in the Human Female*, 129
Kittler, Friedrich, 30–31
Klein, Melanie, 13–14, 58, 140, 152
Knott, Sarah, 18, 26–27
Kristeva, Julia, 18

Labor-saving devices. *See* Time/labor-saving devices
Lacan, Jacques, 16
Ladies' Home Journal (magazine), 91, 92, *123*, 127, 187, 189
Lafargue Clinic, Harlem, New York, 12, 187–190, 192–193
Lamarckianism, 11
Lambs, 119–120
Landefeld & Hatch, 221–222
Laplanche, Jean, 16
Larson, John A., 182
Latchkey children, 200, 282n51
Law, attachment theory's use in, 163, 166–168

Lazarsfeld, Paul F., 22, 185
Leeper v. Leeper (2009), 167
Left, political, 7
Lesbian mothers, 102
Levine, Elana, 198
Levine, Madeline, 143
Levy, David M., 17, 22, 126, 203
  Maternal Overprotection, 126
Lewis, Sophie, 26, 217
Libidinal mothering, 122
Liedloff, Jean, 103, 271n82
LIFE (magazine), 91, *93*, 102
Lindbergh baby kidnapping, 35–38, 60
Little Albert, 53–54
London Child Guidance Clinic, 191

Maciak, Phillip, 31, 179
Macy Foundation, 13, 122
Mahler, Margaret, 16, 17
Main, Thomas, 253n42
Malcolm X, 149
Malin, Brenton J., 182
Malinowski, Bronisław, 37
Mander, Jerry, 212
Marano, Hara Estroff, 143
Marenholtz-Bülow, Baroness Bertha von, 81
Marital roles, 235
Marneffe, Daphne de, 74
Mason, Roy, 229–234
Masturbation, 83
Maternal deprivation. *See also* Cool mothers
  animal studies on, 119–120
  attachment theory and, 151
  and autism, 113–114
  prisons as site for studying, 149–150, 155
  scientific studies on, 102, 113–114, 118–121, 153–154
  technological supplements as contributor to, 94
Maternal environment, 15–16, 48, 55–57, 250n20
Maternal impression, 10, 11, 46, 80, 83, 186

Maternal impulses, 81, 83, 97
Maternal instincts
   advocacy of, 13–14, 57
   criticisms of, 78, 81, 97–98
Maternal matrix, 16
Matriarchy, 129, 130
Matrix concept, 16, 18
May, Elaine Tyler, 26, 129
McCall's (magazine), 102
McClintock, Anne, 58
McDonald, Eugene F., Jr., 60
McLuhan, Marshall, 24, 30, 115–118, *115*, 121, 181, 187, 198, 199, 212
Mead, Margaret, 17, 275n36
Mean world hypothesis, 211
Mechanical Slave, 62
Media-as-mother panics, 200–203
Media contagion, 39, 183–84
Media ecology, 8, 22, 24–25, 181, 212
Media/mediation. *See also* Comic books; Movies; Remediation; Television
   autism linked to, 140–141
   cool, 115–117, 198
   criticisms of, 5–6, 20–22, 180, 243
   family as anecdote to, 1–3
   hot, 115–117
   influence-theories of, 23
   mid-twentieth-century family and, 4–5
   milieu concept in relation to, 16
   mothers/mothering conflated with, 3, 5–10, 15–25, 28, 30–31, 57, 141, 180, 213
   psychology and the concept of, 20
   race and, 180
   reciprocal relations in, 19
   sleep threatened by, 88
   as stand-ins for nurses/nannies, 38–40
Media panics, 179–182, 187
Media studies, 8, 23–24
Media theory, 6, 8, 20–25
Melody, William, 211
Menaker, Esther, 18

Menninger, Karl, 152
Metzl, Jonathan, 133
Meyer, Adolf, 12–13, 15, 17, 127, 187
Michel, Sonya, 41–42
Microaggressions, 206
Microsoft, 234
Middle class
   and childcare crisis, 29
   cool mothering characteristic of, 118, 132–133
   and domestic care, 37–39, 43, 46–48, 53, 64, 74, 179–180
   goods and technology available to, 3
   helicopter parenting characteristic of, 143
   home design for, 222, 226–228, 230, 233
   idealization of, 26, 28, 57, 59, 150
   ideals of motherhood for, 103, 106, 108, 123, 202, 235
   Sesame Street's benefits for, 210
   technological supplements for, 47, 63, 65, 86–87, 94
   telework in, 217
   working women in, 201, 217, 220, 229
Milieu concept, 16
Millard, Herman H., 86, 262n20
Mills, Mara, 274n23
Milner, Marion, 18
Milton, John, 242
Misogyny, 18, 19, 115, 128, 134, 144
Model homes, 218–225, 233–234
Moloney, James Clark, 15, 17, 203
Momism, 127, 130, 143, 226
Monitors. *See* Baby monitors
Monkeys, 100, 114, 119, *120*
Monsanto House of the Future, 223–226, *224*, 230
Montessori schools, 237
Moore, Kelli, 278n66
Moore, Terry Jean, 165–166
Moral contagion, 183–184
Morgan, Jennifer, 148
Morris, Charlotte, 177

Morris, Norman, 211
Morrisett, Lloyd, 205
Morrison, Taylor, 233
Morrow, Robert W., 209
Mosse, Hilde, 188
Mother-baby, 55–57
Mother-child relationship. *See also*
    Attachment theory; Attention and
    affection; Maternal deprivation
  communication in, 125–126
  cool, 113–115, 130–140
  hot, 114–115, 121–131, 269n42
  nannies' interruption of, 46
  prisons as site for, 149–151, 153–160,
    163–173
  severing of, 146–149
  Winnicott's conception of, 55–57
Mothers/mothering. *See also terms*
    beginning with Fitness, for motherhood;
    Housewives; Maternal; Mother-child
    relationship; Parenting/child-rearing;
    Social reproduction; Work, mothers'
  always-on, 5, 29, 143, 218, 235
  "bad," 5, 11, 25, 30, 113–114, 125,
    127–128, 130–131, 139, 160, 180, 184,
    190–191
  Black, 130–131
  cool, 15, 113–115, 130–140
  criticisms of, 4–7, 11, 17, 23, 25, 27, 55,
    121–144, 226
  defense and support of, 122
  democracy linked to, 26, 127–128
  double bind of, 20, 29, 57, 60, 68, 75, 108,
    125, 139, 144
  "good," 5, 16, 19, 122, 158, 163, 166, 171,
    208, 271n82
  good-enough, 55, 120
  hot, 114–115, 121–131, 269n42
  immersion, 104
  importance and responsibility attributed to,
    5, 8, 10–11, 14–16, 19, 30
  incarcerated, 147–173

  influence-theories of, 23
  intensive, 29, 95, 103, 225–226, 228, 235
  libidinal, 122
  material transfers to children from, 9–10,
    19, 24
  meanings of, 26–27
  media/mediation conflated with, 3, 5–10,
    15–25, 28, 30–31, 57, 141, 180, 213
  mid-twentieth-century ideal of, 3
  nannies in relation to, 38, 50, 60
  outside influences on, 21–22
  and postpartum child-rearing problems,
    260n94
  psychological and other social scientific
    study of, 4–19, 113–114
  purity of, 7, 19, 23
  reciprocal relations in, 18–19
  refrigerator, 15, 113–115, 121, 124, 130,
    133–137, 226, 270n65
  remediation of, 6–7, 20–21
  and sexuality, 121, 129
  smother, 124–125, 143
  surveillance of, 69
  and television, 197–198, 201
  tiger, 142
  violence in conceptualization of, 26–27
Motion Picture Research Council, 182, 187
Mount Sinai Hospital, New York, 12, 153
MOVE organization, 243
Movies, 22, 181–186
Mower, Molly, 90
Mower, O. H., 90
Moynihan Report (1965), 130–131, 207,
    269n40
MTV, 200
Mullendore, Richard, 143
Multigenerational families, 3–4
Mumford, Kevin J., 131
Musser, Amber Jamilla, 58

NAACP. *See* National Association for the
    Advancement of Colored People

Nannies/nurses. *See also* Childcare; Domestic workers; Nanny/nurse contagion; Nanny panics; Wet nurses
and class, 59–60, 74–75
contamination threat from, 45–47, 62, 85
double bind of, 59
enslaved, 45
etymology of "nanny," 43
history of, 43–44
media/technology as stand-ins for, 38–40
mothers in relation to, 38, 50, 60
as a problem/threat, 44–48, 50, 64, 67 (*see also* Nanny panics)
professionalization of, 44, 50
psychology and, 51–55, 57–58, 74–75
race of, 45–46, 75
roles of, 43, 59
sexual seduction by, 51–53
surveillance of, 69–75
technological replacement of, 59–63, 66
violence suffered by, 45
Nanny cams, 69–75. *See also* Baby monitors
Nanny/nurse contagion, 43, 47, 73, 85
Nanny panics, 40, 44–46, 53, 59–61, 67–75, 85
NASA, 232
Natal alienation, 27
National Association for the Advancement of Colored People (NAACP), 188
National Comics (DC Comics), 194
National Society for Autistic Children, 136
Nazis, 118, 239
NBC (TV network), 188
Neal, Mary, 101–102
Rock-A-Bye-Baby, 102
Neo-behaviorism, 105
Neurodivergence, 140–141
New woman, 80
New York Foundling Hospital, 100–101
New York Institute of Child Guidance, 126
New York Magazine, 209

New York Psychoanalytic Society and Institute, 17
New York State Reformatory for Women, Bedford, New York, *157*
New York Times (newspaper), 35, 127, 211–212, 242
Ngai, Sianne, 66
Nietzsche, Friedrich, 190
Nixon, Richard, 208–209
Noguchi, Isamu, 60–61, 65
Nooney, Laine, 231
North, Sterling, 186
Northpoint, 146
Nostalgia
and the future, 287n44
for housewife era, 104
for nurse/nanny care, 38–39
remediation linked to, 38–39, 44, 62, 66, 85
for unmediated mothering, 103–104
Nuclear family
contamination of, 44
ignoring of other types in favor of, 57
as mid-twentieth-century ideal, 3, 5, 9, 25, 38, 63, 186
origin of the term, 9, 37–38
purity of, 39
Nursery schools/day nurseries, 59, 63
Nurses. *See* Nannies/nurses; Wet nurses

O'Brien, M. E., 49
Oedipus complex, 38, 51, 53, 123, 129, 136, 207
Office of the Surgeon General, 204
Ogata, Amy, 224
O'Hara, Frank, vii
Olivier, Lawrence, 96
Orphans, 7, 99, 101, 152, 160
Ovesey, Lionel, 269n42
Owlet Smart Sock, 108

Pacification, 20, 48, 80, 83, 84, 104, 179, 180, 196–198, 204, 210, 232, 289n74

Panama-Pacific International Exposition (San Francisco, 1915), 218–219
Pandemic. *See* COVID-19
Panics
  babysitter, 63
  comic book, 186–195
  media, 179–182, 187
  media-as-mother, 200–203
  nanny, 40, 44–46, 53, 59–61, 67–75, 85
  parenting, 66–67
  Satanic, 44–45, 68
  social, 5, 6, 10, 23, 25, 27, 30, 31
  television, 199
Parenting/child-rearing. *See also* Childcare
  attachment, 103–105
  costs of, 28–29
  detachment, 103
  helicopter, 115, 141–144
  intensive, 103
  as labor, 29, 92
  panics over, 66–67 (*see also* Babysitter panics; Nanny panics; Social panics)
  popular books on, 11, 12, 54, 122–124
  rights of parents, 145, 163–167
  technology for, 28–29, 91–94
Parenting literature, 81, 82
Parenting theory, 6, 8. *See also* Child development theory
Park, Robert, 182–184
Parker, Al, illustration for Ray Bradbury's "The World the Children Made," x
Pastore, John, 204
Patton, Elizabeth, 222
Payne Fund Studies, 21, 180, 182–187, 197, 204
Pediatrics
  and the Air Crib, 95–97
  and mothering, 8, 9, 11, 101–102, 104, 113–114, 123
  and sleep practices, 73, 83, 85, 89, 99
Peters, John Durham, 8, 132
Piaget, Jean, 58

Piecework. *See* Home work
Pierce, Chester, 206–208
Pinchevski, Amit, 132
Play, 237
Playpens, 48, 103
Playrooms, 1–2, 221–222, 232
Play therapy, 13
Politics. *See* Democracy
Portwood-Stacer, Laura, 241
Postman, Neil, 24, 212
Postpartum child-rearing, 260n94
Prams, 56–57, 82
Predictive theories, 15, 149–150, 152, 160, 161, 163, 172
Prefabricated houses, 222–223
Primary affect hunger, 126
Prisons/incarceration, 145–173
  attachment theory and, 149–150, 165–172, 279n93
  carceral capitalism, 146–147, 170, 173
  class and, 145–147
  digital algorithms and practices in, 145–147, 150–151, 164
  mothers in, 147–173
  Progressive era theories of, 158
  race and, 145–148
  studies conducted in, 149–151, 153–160
Progressive era, 27, 38, 46–47, 79, 84, 85, 148, 158, 160, 172, 182
Propaganda, 181
Pseudo Programs, 200
Psychoanalysis. *See also* Child psychoanalysis
  fragmenting of field of, 14
  growing importance of, 12
  and nannies, 51–53, 55, 57–58
  study of mothers/mothering by, 12–14, 114, 152
  and television, 252n42
  in the United States, 18
  women theorists' role in, 57–58
Psychoanalytic Research Project on Problems of Infancy, 154

Psychogalvanometers, 182–183, *183*
Psychology. *See also* Child development theory; Psychoanalysis; Scientific studies of children
  and media concept, 20
  and nannies, 51–55, 57–58, 74–75
  and *Sesame Street*, 205–210
  study of mothers/mothering by, 4, 6–9, 12–19
  study of the family by, 2, 7, 9
  toys analyzed from perspective of, 249n4
Psy-ences
  on childcare, 29
  destruction of, 146–149
  feminization of, 17–18
  and incarcerated mothers, 149–151
  and nannies, 51–53
  research on mothering and mediation, 5–6
  white supremacy in, 18
Pulse oximeters, 108–109
Purity
  imputed to children, 241–243
  imputed to the family, 7, 28
  imputed to mothers, 7, 19, 23
  imputed to the nuclear family, 39

Quiet: We Live in Public (social experiment), 283n54

Race/racism. *See also* African Americans; Whites/whiteness
  children's status and, 49
  and fitness of mothers, 9, 172
  incarceration practices and, 145–148
  and media criticism, 180
  in medical system, 133–134
  mothering styles and, 118, 130–131, 134
  and nannies/nurses, 45–46, 75
  and postpartum child-rearing problems, 260n94
  psychoanalysis and, 18
  and sleep practices, 83

Steiner's philosophy and, 239
  television and, 206–207
  violence associated with, 190, 192, 206–207
  Wertham's psychological method and, 187–188
Radio, 181
Radio Corporation of America (RCA), 65, *66*
Rayer, Rosalie Alberta, 53–54
Raz, Mical, 147
Reality TV, 200
Real Womanhood, 27
Real World, The (TV show), 200
Recidivism, 146, 148, 164, 166, 170, 172
Reconstruction, 46
Redbook (magazine), 196, 198
Refrigerator mothers, 15, 113–115, 121, 124, 130, 133–137, 226, 270n65
Reggio schools, 237
Reich, Wilhelm, 38
Remediation
  criticisms of, 23
  medium made visible by, 21
  of mothering, 6–7, 20–21
  nostalgic aspect of, 38–39, 44, 62, 66, 85
  of nurses/nannies, 39–40
  parental fears concerning, 29
Remote work. *See* Telework
Richard, Grace, 209
Rietmann, Felix, 155, 161
Right, political, 7
Rights, parental, 145, 163–167
Roberts, Dorothy, 147, 148
Robots and automata. *See also* Technology
  for child-rearing, 42, 67
  Mechanical Slave, 62
  as substitutes for servants, 39
Rockefeller Foundation, 13, 122, 158
Rocking, 77, 83–84, 86, 102
Roosevelt, Eleanor, 88
Rorschach test, 126, 155, 190
Rose, Nicolas, 151–152
Rosenbaum, Susanna, 108

Rosenthal, Caitlin, 148
Roulet, Elaine, 169, *170*
Roy, Deb, 78

Salk, Jonas, 100
Salk, Lee, 100–103
  What Every Child Would Like His Parents to Know, 102
Satanic panic, 44–45, 68
Scannell, Paddy, 203
Schectman, Anna, 8
Schizophrenia, 117, 125, 132, 133–134, 137
Schools. *See* Education and schools
Schreber, Daniel Paul, 137
Schwartz, Laura, 72
Schwem, John W., 219
Scientific mothering
  authority imputed to, 11
  decline of, 13
  defined, 78
  emergence of, 15, 78
  on family schedules, 88–89
  and Freudian psychology, 123
  neo-, 78–79, 95, 103–110
  and sleep, 82–83
  Spock and, 97–98
  and surveillance of mothers, 11, 47
  technological supplements used in, 78–79
Scientific studies of children
  Air Crib used for, 97
  behavioral conditioning, 53–54, 114
  important sites for, 12, 99–101, 118, 153–154
  on incarcerated mothers, 149–151, 153–160
  on maternal deprivation, 102, 113–114
  media used in, 161–162
  movies as object of, 182–186
  on sensory factors, 100–101
  on separation of children from parents, 7, 118–121, 153 (*see also* on maternal deprivation)
  sleep-related, 88–90
  television as object of, 196–197
Sconce, Jeffrey, 250n9
Scott, Daryl Michael, 130, 268n40
Screen parking, 179, 180
Screen time
  anxiety about, 179
  coining of the term, 23, 179
  as component of mothering, 29
  cost-benefit analysis for, 179
  criticisms of, 23
Sears, 90
Sears, Martha, 104
Sears, William, 104–105
Securitone Heartbeat Comforter, 100–101
Senate Subcommittee to Investigate Juvenile Delinquency, 194–195
Sendak, Maurice, 203, 205
Sentencing guidelines, 145–147, 164
Sentencing Reform Act (1984), 164
Servants. *See* Domestic workers; Nannies/nurses
Sesame Street (TV show), 17, 24, 204–211
Sexuality. *See also* Homosexuality
  mothers and, 121, 129
  nannies/nurses and, 51–53
Shaken baby syndrome, 70–73
Sharp, Violet, 35–36
Sheftel, Susan, 74
Short, William H., 182–183
Show homes. *See* Model homes
Sibling rivalry, 126
SIDS. *See* Smothering
Silverman, Chloe, 137
Siri (virtual assistant), 230
Skinner, B. F., 91–100
Skinner, Deborah, 91–92, 94, 263n36, 264n49
Skinner, Yvonne, 91
Skinner boxes, 91, 264n49
Skin-to-skin contact, 103–104

Sleep. *See also* Co-sleeping
  characterological effects of, 88
  as component of a healthy family, 78, 82, 95
  independent, 77, 82, 84–86, 95
  mediated and unmediated, 79–80
  problems/threats associated with, 79, 82, 88
  social meanings attached to, 82
  technology for, 77–78, 80, 86–88, *87*, 90–95
  tracking and alerting technology for, 108–109
  working parents and, 78, 82, 94, 107–108
Sleep position, 82, 98–99, 107
Sleep training, 105–107
Smart homes
  and childcare, 217, 221–222, 225, 227–229, 232, 235
  computers and, 227–229
  as a media-mother, 216, 218
  model homes, 218–225, 229–234
  and mothers' work, 216, 229, 231–232
  origins of, 9, 218–219
  personal and social effects of, 1, 217
  and whiteness, 225, 235
  Xanadu and, 229–234
Smart House (film), 289n72
Smothering, 83–84, 91, 97–99, 107
Smother mothers, 124–125, 143
Snoo Smart Sleeper, 77–78, 107–108
Snyder, David, 94, 222, 224
Social contagion, 186, 192
Social panics, 5, 6, 10, 23, 25, 27, 30, 31
Social reproduction
  feminist psychoanalysis and, 140
  mothering conflated with, 15, 19
  older forms of, 5
  sleep and, 79, 110
  technology for, 47, 218
Sonia Shankman Orthogenic School, Chicago, 136–138
Sonotone, 100

Sound, 100–101
Spanish flu, 237, 242
Spigel, Lynn, 8, 23, 25, 44, 65, 129, 180, 195, 197, 202, 231, 287n44
Spillers, Hortense, 26, 131
Spitz, René, 15, 119, 122, 142, 150, 153–162, 165, 171–173, 275n36
Spock, Benjamin, 12, 13, 17, 22, 54, 55, 64, 97–99, 103, 122–123, 203, 208, 271n82
  Common Sense Book of Baby and Child Care, The, 11, 12, 64, 97–99, 122
Springhill Stock Farms, Greenwood, Indiana, 119
Springle, Herbert A., 209–210
Stability, familial, 145, 147–148, 157
Starosielski, Nicole, 116
Staudenmaier, Peter, 239
Stearns, Peter N., 82
Steedman, Carolyn, 52
Steinberg, Jacqueline, 166
Steiner, Rudolf, 237–243
Step-Saving Kitchen, A (documentary), 227
Stern, Daniel, 18
Stevenson, Brenda, 148
Stewart, Catherine, 62
Stimuli
  conditioning of, 54
  media as source of, 88, 116–117, 180, 199
  mothers as source of, 20, 114–115, 117, 119–121, 124, 126, 130, 134, 137, 139, 141–144, 226
Stockton, Kathleen, 131
Stoltzfus, Emilie, 63
Stowe, Harriet Beecher, 48
Strange Situation test, 161–162, 168, 171
Strecker, Edward, 126–129, 133, 140, 143, 144
  Her Mother's Daughters, 128–129
  Their Mother's Sons ("The Mom Lecture"), 126–127
Strecker, Gillian, 58
Strollers, 28

Suburban homes/suburbia, 4, 64, 66, 115, 134, 137, 141, 205, 222–227
Sudden Infant Death Syndrome. *See* Smothering
Summers, Martin, 18
Sunrise lamps, 107
Superman (comic book), 190
Surrogacy, 102
Surrogate humans, 284n5
Surveillance, by helicopter parents, 144
Surveillance technologies. *See also* Baby monitors
   childcare workers subject to, 72
   children subject to, 40, 63–66, *66*, 74
   mothers subject to, 69
   nannies subject to, 69–75
   scientific mothering and, 11, 47
   television and, 200–201
Sutherland, Tonia, 157
Szonzi test, 155

Taylorism, 86, 227
TBO Tech, 72
Technology. *See also* Nanny cams; Robots and automata
   addiction to, 2
   criticisms of, 92–93, 96–97
   feminism and, 255n19
   for housewives, 91–92, 106, 135–136, 219, 223–229
   meanings attached to, 202
   and the middle class, 3, 47, 63, 65, 86–87, 94
   in mid-twentieth-century homes, 3
   mothers/parents supplanted and alienated by, 1–3, 73
   nannies supplanted by, 59–63
   for parenting/child-rearing, 28–29, 91–94
   pediatrics and, 96–97
   scientific mothering's use of, 78–79
   sleep-related, 77–78, 80, 86–88, *87*, 90–95
   as stand-in for nurses/nannies, 38–40, 65

Steiner's philosophy and, 237–243
time-saving rationale for, 38, 59, 62, 66, 92, 94, 218–220, 222–223
Television. *See also* Closed-circuit television
   benefits imputed to, 179–180, 195–197, 204–211
   criticisms of, 23, 195–205, 209–212
   educational uses of, 204–211
   in lamb-rearing study, 119–120
   meanings attached to, 195
   mothers and, 197–198, 201
   popularity of, 199
   as prescriptive technology, 177–178
   psychoanalysis and, 252n42
   public policy and, 17, 23, 203–204
   race and, 206–207
Television panics, 199, 201, 203
Telework
   contradictions of, 215–216
   gendering of, 231–232
   smart homes and, 226, 229, 231–232
Temperature, of mother-child relationships, 115, 117
Terry, Jennifer, 129
Test patterns, television, 178–179, 184
Theweleit, Klaus, 30
Thibeault, Jennifer, 72
Thompson, Jean A., 192
Tiger mothers, 142
Tilley, Carol, 192
Time (magazine), 61, 113
Time/labor-saving devices, 38, 59, 62, 66, 92, 94, 135, 218–220, 222–223
Today (TV show), 102
Tomorrowland, Disneyland, 223
Topper, Robert, 96
Torok, Maria, 140
"Town of Tomorrow," 221
Toys
   for infants and toddlers, 237
   optical, 21, 202
   psychological analysis of, 249n4

Trow, George W. S., 212–213
Turner, Fred, 8, 26, 274n23

Umbilical cord, 128, 139, 143
University of Wisconsin–Madison, 119

Vaccines, 238, 240–241, 243
Variety (magazine), 208
Vicedo, Marga, 152
Victorian culture, 3, 40, 51–52
Videotape, 161–162, 168
Violence
   comic-book, 189–191
   familial, 151, 191
   media, 206, 211
   racial, 190, 192, 206–207
   television, 198, 203–204
Vora, Kalindi, 39, 284n5

Wajcman, Judy, 226, 255n19
Waldorf-Astoria Cigarette factory, 239
Waldorf schools, 237–243
Wallerstein, Robert S., 252n42
Walt Disney Imagineering, 223
Wang, Jackie, 146, 274n23
Warhol, Andy, 178
Watch with Mother (TV show), 204
Watson, John, 11, 13, 53–54, 83, 89–90, 95, 109, 123
   Psychological Care of Infant and Child, 11, 54, 89–90, 114
Watter, Seth Barry, 163
Weber, Samuel, 242
"Wee-Alert," 90, *90*
Weiner, Lynn, 46–47
Weinstein, Deborah, 128, 206, 274n23
Weitzenkorn, Rachel, 154, 156
Welfare queens, 180
Wertham, Fredric, 17, 22, 187–195, *189*, 197–198, 203
   Seduction of the Innocent, 189, 191, 194, 195, 197

Wessely, Christina, 16
Westfield State Farm Reformatory, Bedford, New York, 156, 158–159, 169
Westinghouse, 227
Westinghouse Research Laboratories, 62
Wet nurses, 10, 45–46
White, Simone, vii
White House Conference on Children (1970), 102
White House Conference on Children and Youth (1950), 17, 22, 203
Whites/whiteness
   autism associated with, 113, 132–134
   childcare associated with, 28
   and children's status, 49–50
   as cool mothers, 130, 134
   domestic ideal of, 44
   family dynamics of, 131
   helicopter parenting and, 142
   Lindbergh kidnapping and, 36–37
   as norm in psychoanalysis, 18
   smart homes and, 225, 235
   Steiner's philosophy and, 239
   and technological substitutes for labor, 62
   television and, 206–207
Wiener, Norbert, *Cybernetics*, 128
Williams, Raymond, 3
Winn, Marie, 211–212
Winnicott, D. W., 15, 19, 24, 55–58, 64, 105, 120, 122, 152–153, 158, 250n20, 251n29, 275n28
Wisconsin, 145–146, 163, 166
WNEW-TV, 177–178
Wolf, Katherine, 154–155
Women's Liberation, 68, 70, 228, 234
Woodward, Louise, 70–71
Woolf, Virginia, 217
Wordsworth, William, 83
Work, mothers'. *See also* Telework
   barriers to, 63
   childcare conducted simultaneously with, 41–42, 86–87

Work, mothers' (cont.)
  and cool mothering styles, 135
  criticisms of, 71, 202
  double bind of, 108, 135, 215–216
  historical statistics on, 220, 228, 229, 230
  prevalence of, 68
  sleep of babies linked to, 78, 82, 94
World's Fair (Chicago, 1933), 221
World's Fair (New York, 1939), 221
World War II, 6, 7, 12, 118, 121, 152, 159, 222, 282n51
Wright, Richard, 188
Wylie, Philip, 127, 143
  Generation of Vipers, A, 127

Xanadu (Computerized Home of Tomorrow), 229–234, *233*

Yale University Clinic of Child Development, 11, 12, 16
Ye'kuana people, 103

Zelizer, Viviana, 10, 49, 81
Zenith, 60–62